SpringerBriefs in Philosophy

SpringerBriefs present concise summaries of cutting-edge research and practical applications across a wide spectrum of fields. Featuring compact volumes of 50 to 125 pages, the series covers a range of content from professional to academic. Typical topics might include:

- A timely report of state-of-the art analytical techniques
- A bridge between new research results, as published in journal articles, and a contextual literature review
- A snapshot of a hot or emerging topic
- An in-depth case study or clinical example
- A presentation of core concepts that students must understand in order to make independent contributions

SpringerBriefs in Philosophy cover a broad range of philosophical fields including: Philosophy of Science, Logic, Non-Western Thinking and Western Philosophy. We also consider biographies, full or partial, of key thinkers and pioneers.

SpringerBriefs are characterized by fast, global electronic dissemination, standard publishing contracts, standardized manuscript preparation and formatting guidelines, and expedited production schedules. Both solicited and unsolicited manuscripts are considered for publication in the SpringerBriefs in Philosophy series. Potential authors are warmly invited to complete and submit the Briefs Author Proposal form. All projects will be submitted to editorial review by external advisors.

SpringerBriefs are characterized by expedited production schedules with the aim for publication 8 to 12 weeks after acceptance and fast, global electronic dissemination through our online platform SpringerLink. The standard concise author contracts guarantee that

- an individual ISBN is assigned to each manuscript
- each manuscript is copyrighted in the name of the author
- the author retains the right to post the pre-publication version on his/her website or that of his/her institution.

More information about this series at http://www.springer.com/series/10082

Shane Epting

Saving Cities

A Taxonomy of Urban Technologies

 Springer

Shane Epting
Missouri University of Science and Technology
Rolla, MO, USA

.

ISSN 2211-4548 ISSN 2211-4556 (electronic)
SpringerBriefs in Philosophy
ISBN 978-3-030-85832-2 ISBN 978-3-030-85833-9 (eBook)
https://doi.org/10.1007/978-3-030-85833-9

This Springer imprint is published by the registered company Springer Nature Switzerland AG
The registered company address is: Gewerbestrasse 11, 6330 Cham, Switzerland

Acknowledgements

Sections of Chaps. 2–4 first appeared in Epting, S (2020) Infrastructure, Urban Sprawl, and Naturally-Occurring Asbestos: An Ontological Thought Model for Wicked and Saving Technologies. *Open Philosophy*, *3*(1): 389–399, https://doi.org/ 10.1515/opphil-2020-0133. This book is dedicated to Ana.

I thank the philosophers I have hung out with at many of the Philosophy of the City conferences over the years. They are listed below in no particular order.

Michael Menser, Jules Simon, Michael Nagenborg, Sanna Lehtinen Ronald Sundstrom, Judith Green, David Woods, Samantha Noll, Daniel Weinstock, Avner de-Shalit, Kwabena Edusei, Gray Kochhar-Lindgren, Joey Tuminello, Diane Michelfelder, Paul Thompson, Cynthia Willett, Jason Matteson, Paula Cristina Pereira, Taylor Stone, John Kaiser Ortiz, Osiris González, Karen Wall, Brit Schulte, Matthew Crippen, Luis Ruben Diaz Cepeda, Víctor Manuel Hernández Márquez, Adeola Enigbokan, Birge Yildirim, Katherine Davies, Joey Aloi, Ilana Boltvinik, Achille Varzi, Eduardo Mendieta, Daniel Bell, Frank Cunningham, Lewis Gordon, Brian Elliott, Roger Paden, Remmon Barbaza, Gerald Erion, Nir Barak, Enrique Dussel, Fábio Valenti Possamai, Gehad Abdelal, Abel Franco, Mary Carmen Marcous, Sarah Gelbard, Dominique Lämmli, Kenny Easwaran, Tyler Zimmer, Arthur Biermann, Roman Meinhold, Magali Bessone, Hernando Estévez, Vesa Vihanninjoki, Arto Haapala, Francisco Colom González, Lauri Lahikainen, Melissa Cardenas, Ferenc Hörcher, Yuiza Martínez Rivera, Bart van Leeuwen, Michael Merry, Pieter Vermaas, Claire Revol, Vladan Klement, Sharon Stanley, Joel Michael Reynolds, Daniel Restrepo, Bri Gauger, Nicola Sidi, Edgar Eslava, Brandt Dainow, Margoth González Woge, and Tea Lobo (even though we've only met online). Thanks to Christopher Coughlin, Anita Rachmat, Arumugam Deivasigamani, and Ties Nijssen at Springer.

Contents

About the Author

Shane Epting is an Assistant Professor of Philosophy at the Missouri University of Science and Technology. He is a cofounder and Codirector of the Philosophy of the City Research Group. His research interests include infrastructure, transportation, food systems, and future cities. He is the author of the book, *The Morality of Urban Mobility: Technology and Philosophy of the City.*

Chapter 1
Introduction

Abstract This chapter is an introduction to the book. It provides an overview of the following chapters and introduces the main ideas and themes of the investigation. This introduction familiarizes readers with the primary concepts, briefly mentioning how they fit together to advance a coherent line of thought. These topics include climate change,wicked technologies, saving technologies, mental life, and saving cities.

Keywords Introduction · Ideas · Themes · Topics

Several technologies found in cities contribute to wicked problems such as climate change and injustice, helping create unfavorable conditions for mental life. I refer to these devices as "wicked technologies." Cities, as technologies, can be wicked as well. They need saving. The path to salvation requires removing, replacing, and creating urban technologies that mitigate harm and support worthwhile goals such as socially just sustainability and human flourishing. Such technologies, considering that they can save us from an anthropogenic demise, count as saving technologies. The book's purpose is to illustrate these points. This introduction touches on the chapters' themes and describes the methodology employed for their execution.

For instance, in Chap. 2, Wicked Technologies, I examine how the conception of "wicked problems" helps us understand the challenges found when cities expand. I use an example from a construction site in southern Nevada, where newly discovered deposits of naturally occurring asbestos halted crews from building a roadway. This case serves as a "thought model," challenging a well-known way of categorizing technologies. Enhancing studies in this area requires developing new categorizations that account for a different way of thinking about how urban technologies affect people, zeroing in on mental life.

Chapter 3, Cities and Mental Life, looks at how urban technologies play roles in causing mental states such as anxiety, fear, stress, and depression. Such situations have a lengthy history, and confronting it is an initial step in its mitigation. Through expanding our taxonomy to include wicked technologies, along with a new term "saving technologies," we can employ these expressions to clarify human-technology relations in cities, focusing on specific cases that produce harmful mental states for urban residents. Some examples include homelessness, traffic, and social isolationism. Due to familiarity, some people could view those realities as cities' inherent qualities—a sadly mistaken view. Seeing them as circumstantial alters our perception of cities' possibilities, leading to different (and hopefully better) urban futures. This chapter grounds this undertaking, identifying some wicked technologies and calling for a dedicated study to develop mitigation.

Chapter 4, Saving Technologies, further expands the taxonomy of technologies, showing some of their possible impacts on mental life. It illustrates that moving away from such situations requires a new conceptual model that can secure mitigatory efforts and alternatives that favor the worthwhile goals mentioned above. In brief, these are saving technologies. This chapter explores them, revealing the motivations for their acceptance and embrace.

By understanding saving technologies, Chap. 5, Saving Cities, examines the benefit of thinking about cities as technologies, mainly saving ones. It shows how they can help increase positive mental states such as joy and love, which can help create better mental lives for urban residents. It exhibits how the structure of "saving" can facilitate immediate efforts that can help shape mental life in the city. To clarify these terms, this chapter explores a novel housing technology: cohousing. By using this example, a picture begins to come into focus, showing how saving technologies could have a significant impact with meaningful research, attention, and support.

Chapter 6, Saving Cities: The Road Ahead, identifies and explores nascent alternatives that could help municipalities gain saving-city status. It shows what a saving city could look like. A comprehensive picture comes into focus by exploring additional saving technologies such as aerial cable cars, participatory budgeting, and car-free cities. Although these technologies will require significant research and development, examining their conceptual images provide motivation and guidance for their possible place in future cities. They also give a clearer picture of what saving technologies and saving cities look like.

In sum, this book has the clear primary goal of expanding the taxonomy of technologies to account for the urban condition, an ontological adventure. An epiphenomenon arising from this exploration is that the research methodology must change to see the significance of this approach requires venturing outside of customary philosophical inquiry. Rather than engaging exclusively with philosophers and philosophically inclined thinkers, it pushes away from what Lewis Gordon (2015) calls 'disciplinary decadence,' wherein philosophy, as an academic discipline, is the superior form of investigation; all others are beneath it. Instead, it views philosophy as one research device that works best in conjunction with the discoveries from its academic neighbors. It moves towards what Gordon (2008, p. 155) calls the "teleological suspensions of disciplinarity." As Gordon (2008, p. 155) puts it: "This involves keeping in sight the purpose of our reflections and an attunement to reality. In philosophy, this involves a teleological suspension of philosophy. It is when philosophers do not worry about being recognized as philosophers; it is when they do not expect philosophy to be the answer to every problem."

In turn, the significance that my approach in this book offers is being goal-oriented. One outcome that emerges is not simply producing more philosophy for its own sake or taking on conventional tasks such as filling gaps in the literature. This idea entails that the methodology and correlative literature review are thematic, bringing in several disciplinary contributions to enhance our understanding of familiar situations encountered daily. However, this project remains inherently philosophical. It takes standard urban technologies and fleshes out aspects that reveal something extraordinary, a reality missing in our everyday views of city life. Still, the extraordinary is not absolute, and its significance should not be extremely elevated. However, the philosophical dimensions (of urban technologies) should not be discounted or put beneath the ordinary (aspects of urban technologies). They are extraordinary because they could easily be missed, failing to become appreciated, which is tragic. These aspects enhance the significance of urban life and the city. Yet, they must correlate to the ordinary symbiotically. There can be nothing extraordinary without the ordinary, and the latter remains under-investigated if we ignore or neglect the former.

Bearing these points in mind, it should be no surprise that the research in this book has different objectives than an archetypal philosophical enterprise. The goal is to philosophize about cities and urban technologies that help us take it to the street, the neighborhood, and the city. This positioning lets us examine unpleasant realities about today's municipalities. Strategically reconceptualizing them can provide the means to save us from the dire situations that accompany many metropolitan areas. Thinking about urban places geared in this direction can rescue humanity from imperilment. In an alternative sense, it explains the title, *Saving Cities*.

References

Gordon L (2008) Not always enslaved, yet not quite free: Philosophical challenges from the underside of the new world. Philosophia 36(2):151–166

Gordon L (2015) Disciplinary decadence: living thought in trying times. Routledge, New York

Chapter 2
Wicked Technologies

Abstract The conception of "wicked problems" can help us understand the challenges associated with many urban issues that exist now or will arise while building, rebuilding, or expanding cities. This chapter uses a recent case study in southern Nevada to grasp the latter's complexity. For instance, geologists discovered new deposits of naturally occurring asbestos, microscopic fibers found in rocks and soil. The danger is that inhaling them can lead to mesothelioma. One problem is that this rare cancer often takes decades to manifest. This discovery abruptly stalled a highway project near Las Vegas. In turn, I use this case as a "thought model" to challenge an established way of categorizing kinds of technologies. Advancing this conversation requires that we reclassify some of them and develop a categorization for those that reflect a different way of thinking about how urban technologies affect mental life.

Keywords Revealing · Enframing · Wicked problems · Saving power

2.1 Introduction

Cities are wicked to the extreme, meaning they are highly complex, entail uncertainty, and enmeshed in socio-eco systems—to say the least.[1] Science and engineering cannot tame them alone. They need philosophy of the city because it teaches us about their ethical, aesthetic, and ontological dimensions. This chapter concerns the latter. Specifically, it deals with the different categories of urban technologies. Michael Nagenborg (2018, p. 346) defines them in the following way: "'urban technology' refers to technologies that are shaping or are being shaped by city life." Specifically, the focus is on how cities shape people's experiences of the city, how they think about them, and what they mean for urban futures. Considering that humankind will continue to build, rebuild, and expand cities, advanced studies need to keep pace in the areas mentioned above.

---

[1] Kathryn Davies' et al. description here is fitting. For more information, see Davies et al. (2015).

© The Author(s), under exclusive license to Springer Nature Switzerland AG 2021   5
S. Epting, *Saving Cities*, SpringerBriefs in Philosophy,
https://doi.org/10.1007/978-3-030-85833-9_2

For instance, researchers hold that urban centers will appropriate over a million square kilometers of undeveloped land by 2030 (Seto et al. 2012). Numerous infrastructures are required before people can populate these future places, and despite our advanced knowledge of the planet, unknown and unforeseen obstacles remain. One such challenge that developers, planners, and engineers must address is the possibility of encountering unidentified hazards such as naturally occurring asbestos (NOAs) and microscopic fibers present in rocks and soil (Perkins et al. 2008). When construction-related activities such as rock crushing release these fibers into the air, people are at risk of inhaling them (Perkins et al. 2008). The danger is that such exposures can make them suspectable to illnesses such as mesothelioma and lung cancer, conditions that can take decades to manifest (Manning et al. 2002).

Construction crews recently encountered this situation near Las Vegas while building a highway bypass (McCrea 2014). The discovery of NOAs brought production to a halt, requiring management to create safety protocols to ensure that workers were not significantly put in harm's way (McCrea 2014). These measures decreased overall exposures, supporting efforts to keep laborers safe (Akers 2017). Although this issue raises several ethical worries for workplace safety and future residents, there are also ontological concerns that require attention. In turn, exploring this case—specifically the nexus of infrastructure, urban sprawl, and NOAs—serves as a "thought model." It helps us see why technologies that can exacerbate or mitigate environmental and or social harms require a new classification because of how they affect humankind's existence in ways that earlier conceptions did not.

To situate such a category, we need to recall Martin Heidegger's established taxonomy of technology. This new kind makes sense when we think about it in terms of Heidegger's groupings of technologies as "revealing" or "enframing" (Heidegger 1997, pp. 319–325). The name that I have in mind is "wicked technologies," which, same as a wicked problem, does not mean that they are inherently bad. Instead, they require such a title because they are challenging to understand and reveal more significant problems (among other aspects). This notion is consistent with the term "wicked problem," as found in several disciplinary and interdisciplinary works of literature (Davies et al. 2015; Rittel & Webber 1973).

Unpacking the claims above is the purpose of this paper. It begins by reviewing the term "wicked problem." Next, I examine Heidegger's thinking in "The Question Concerning Technology," explicitly paying attention to the terms revealing and enframing. With these conceptions in mind, I examine how emerging infrastructure cases, such as the one mentioned above, reveal the required conditions for wicked technologies. This chapter ends by looking at some possible challenges and sets the stage for subsequent chapters that make use of this expanded taxonomy.

2.2   Wicked Problems

Rittel and Webber (1973) conceptualized the term "wicked problems" to address the complicated nature of planning and policy affairs. Broadly, they put forth ten characteristics associated with such issues. Here is a summation (Peters 2017, p. 388):

> (1) Wicked problems are difficult to define. There is no definite formulation. (2) Wicked problems have no stopping rule. (3) Solutions to wicked problems are not true or false, but good or bad. (4) There is no immediate or ultimate test for solutions (5) All attempts to solutions have effects that may not be reversible or forgettable. (6) These problems have no clear solution, and perhaps not even a set of possible solutions. (7) Every wicked problem is essentially unique. (8) Every wicked problem may be a symptom of another problem. (9) There are multiple explanations for the wicked problem. (10) The planner (policy-maker) has no right to be wrong.

Their formulation explains these points in some detail, but the overall gist is that dealing with these problems requires different or additional skills from scientific ones. This view also suggests that "solutions" cannot amount to much more than fleeting mitigatory efforts, a notion that clearly differentiates tame and wicked problems. Due to the encompassing yet unique character of the kind of troubles that we find in and with cities, using "wicked problems" as a conceptual device is an excellent tool for the job.

Consider, for instance, that today's cities are parts of larger wicked problems such as climate change and vast economic inequalities. Yet, smaller wicked problems such as transportation systems remain part of them. We need a technology of thought that can examine how efforts of varying sizes and kinds will impact a given city to mitigate harm or qualify as a solution—even though using the latter term means defining its conditions as possibly fleeting. Suppose climate change, cities, and certain urban technologies all align with the description above. In that case, its compatibility, even in a gross sense, shows reasons why it is an appropriate tool for the job of helping us think through how we find ourselves situated with climate change, cities, and several urban technologies. In turn, the benefit is that we can learn how a particular dimension of why a specific device causes trouble. We know that we need extra-scientific thinking to approach the problem to include necessary elements for mitigation and its precarious nature. For example, Rittel and Webber (1973, p. 163) make this point evident:

> With wicked planning problems, however, every implemented solution is consequential. It leaves "traces" that cannot be undone. One cannot build a freeway to see how it works, and then easily correct it after unsatisfactory performance. Large public-works are effectively irreversible, and the consequences they generate have long half-lives. Many people's lives will have been irreversibly influenced, and large amounts of money will have been spent-another irreversible act. The same happens with most other large-scale public works and with virtually all public-service programs.

The passage above is significant because it shows the reality of implementing infrastructures that many people use without fully grasping what it means for them to be a part of their cityscape; how it shapes their cities and their lives. For instance, highways can divide communities (Payne 2015). They can harm public health in myriad ways under certain conditions (Srinivasan et al. 2003). Light rail stops can increase property values, which play a role in displacing vulnerable and marginalized residents (Scott 2012). Poorly maintained water resources could poison people, as the case of Flint, Michigan, makes evident (Dana and Tuerkheimer 2016).

These issues raise concerns for science and engineering when formulating proper responses, but knowledge from those areas requires insights beyond their disciplinary boundaries. Herein we find the appeal and utility of "wicked problems" as a conceptual device. Wicked does not exclude the tame, but it acknowledges and solicits the benefit of troubleshooting measures that cannot exist within strictly determined parameters. In turn, venturing into realms such as ethics and ontology can yield insights to make sense of such situations.

The added advantage is that, as a conceptual device, "wicked problems" are congruent with neighboring concepts. Being able to stack and or merge such conceptions expands these devices' usefulness. It exhibits how we can discover additional insights into how we can mitigate harmful effects associated with wicked problems. To illustrate how such a process works when dealing with these affairs, the following section shows how we can bring the notion of wicked problems into the existing taxonomy of technologies to provide insights into human-technology relations.

2.3   Heidegger's Conceptions of Technology

In "The Question Concerning Technology," Heidegger examines technology to learn about aspects of humanity that tell us about being, paying careful attention to how we relate to nonhuman nature through it (Heidegger 1997).[2] He argues that such relations say more about us than they do about technology, warning that the kind of thinking that comes with pursuing technology for its own sake should not be the only way we think (Heidegger 1997). To elucidate this point, Heidegger puts technology into the categories mentioned during the outset.

Revealing technologies are the kind that we associate with ancient societies, in the sense that they "reveal" the power that lies hidden in the Earth (Heidegger 1997, p. 319). The horse and plow and the windmill serve as examples (Heidegger 1997). The former reveals how agricultural practices can yield food, and the latter can provide the power to pump water from the ground (*ibid.*). By categorizing these devices, he shows us the pattern underlying all technologies with such abilities that do not go much beyond this degree of sophistication. He exposes something beneficial that, with some ingenuity and elbow grease, provides humankind with a way to work

---

[2] Other philosophers and technology scholars have works that deal with classifications of technology that could benefit understanding the totality of human-technology relations (e.gs., Winner 1980; Collingridge 1982). However, Heidegger's taxonomy of technology intersects with being, which is closely related mental life as explained in subsequent chapters. Due to this feature, I focus on this ontology exclusively.

with whatever is available. In a rudimentary sense, the description above is the limit for technologies that fit in this category.

However, modern people have numerous devices that exceed the depiction above. Heidegger illustrates why humankind requires another category that accounts for the differences (*ibid.*). He refers to them as enframing (*ibid.*). The central idea behind this concept is that while these technologies reveal the hidden power that ingenuity and the Earth's resources can provide, they do so in a manner that holds natural resources as standing reserves (*ibid.*). This idea is significant because it is the essential element that separates enframing technologies from revealing ones. Consider, for example, how Heidegger compares how a hydroelectric dam has a different impact on the Rhine river than an old, wooden bridge does. For instance, Heidegger (1997, p. 321) holds:

> The hydroelectric plant is not built into the Rhine River as was the old wooden bridge that joined bank with bank for hundreds of years. Rather the river is dammed up into the power plant. What the river is now, namely, a water power supplier, derives from out of the essence of the power station. Whatever is ordered about in this way has its own standing. We call it the standing-reserve [Bestand][3]

The comparison above illustrates how each technology affects the river. It shows why the dam deserves the classification of enframing, considering that rivers become a standing reserve for us to provide electricity (*ibid.*).[4] The river is no longer allowed to only be for its own sake, but it can do so only in the sense that it remains under humankind's control for our purposes (*ibid.*).

His primary concern is not to focus on technology to reveal something about it per se or on how it harms nonhuman life but to study it to form a free relationship to such technologies (*ibid.*). Yet, we must keep in mind that the motivation that drives this "free relationship" is how such relations affect humans. It shows how we relate in general, which implies who we are and how we view the world surrounding us. The idea is that repeating this pattern of behavior across many planes of existence can alter how we think and, in turn, change the conditions for being. For instance, as Hubert Dreyfus (1997) reads Heidegger, the bigger picture to see here is not that Heidegger is retreating from technology, but he holds this position to show that there is a danger wherein we could lose the ability to look at being in any other way. If humankind can develop such a relationship that does not view the nonhuman world as standing reserve, then perhaps a case can be made that we would also not see each other in such a manner.

Yet, when we think about these ideas in the present context of wicked problems such as climate change, we discover that we are facing new challenges. It is a threat that waits for us in the future most fully. However, forces such as hurricanes and extreme weather events already indicate that such concerns are now conditions that impact our being, a point that Irwin (2010) has developed robustly.[5] Although

---

[3] *Ibid.*, 321.

[4] See footnote 3.

[5] What also deserves mention is that this project also, very broadly, fits in with the larger body of scholarship that examines the interplay of studies of Heidegger, technology, and climate change.

Heidegger has lots to say about anxiety and being, I want to go in somewhat of a different direction, employing anxiety and other terms to zero in on mental states and mental life in the city.[6]

I employ the term "mental states" in a non-technical way. I see them as helping compose one's mental life. One could argue that "mental life" is outdated, suggesting the use of "mental health" instead. The goal here goes beyond mental health. The scope of the desired elucidation is much more encompassing. That is, philosophy, in many ways, deals with examining mental life and its numerous dimensions. Mental health has a significant role in a person's mental life, but mental life includes elements such as experiencing mental states, having meaningful conversations and friendships, appreciating art and artifacts, being creative, having dreams, being fulfilled, and having feelings connected to spirituality—if it aligns with one's 'metaphysical' disposition. This point does not entail that I will not use psychologists' insights, only that I do not make any challenges to how they conduct their affairs.

For instance, psychologists now show that climate change is inducing anxiety in several ways (Clayton 2020). Farmers must directly deal with the harmful mental states that climate change brings, which in some cases could increase suicide rates (Berry et al. 2010). Viewed in these terms, we can show how climate change induces anxiety, fear, stress, and depression. In turn, this scenario promotes a proclivity towards these mental states for some people, shaping their existence. By examining this condition, we can extrapolate insights into how urban technologies, regarding wicked problems, can count as wicked technologies, also.

The section below shows how developing this classification is a necessary move that gives us the means to focus on specific technologies with essential conditions that exceed Heidegger's conceptions. To flesh out this view, I examine the nexus of infrastructure, urban sprawl, and NOAs, a thought model that qualifies infrastructures in such scenarios as wicked technologies. In turn, we see how this additional category can help us form a relationship to technology while contending with considerations such as climate change's ill-effects, one that could benefit humankind.

For instance, a few examples of this work, See Irwin, Ruth. *Heidegger Politics and Climate Change.* Also, Stefanovic, Ingrid. *Safeguarding our Common Future: Rethinking Sustainable Development.* To take this line of thought in a different, but a related direction, see Brockelman, Thomas. *Zizek and Heidegger: The Question Concerning Techno-Capitalism.* Although this paper topically fits in broadly with the works above, its goal to expand the taxonomy in a manner that moves towards practical application on a smaller, more manageable scale. For this reason, the municipality makes such endeavors feasible, which counts as a venture into largely unchartered territory in the literature.

[6] For an in-depth examination of Heidegger, anxiety, and climate change that addresses several intersecting and significant philosophical issues, see Myers (2014) "Understanding climate change as an existential threat: confronting climate denial as a challenge to climate ethics." *De Ethica. A Journal of Philosophical, Theological and Applied Ethics* 1(1): 53–70. Elsewhere, I explore this topic in relation to Heidegger and anxiety. See Epting (2020) Infrastructure, Urban Sprawl, and Naturally-Occurring Asbestos: An Ontological Thought Model for Wicked and Saving Technologies. *Open Philosophy,* 3(1): 389–399.

2.4 Wicked Technologies in Focus

Examining how to categorize technologies is somewhat sophisticated. It involves researching abstract concepts while using real-life examples such as named technologies. This approach requires looking at the history of technological studies, an inherently interdisciplinary affair. Such an approach leans towards being integrative. It brings several kinds of areas together to enhance our understanding of the interplay between these topics. The example below also remains consistent with this approach. It involves investigating the abstract subject of categorization with environmental dangers' concrete reality, evident through possible exposures to NOAs. The goal of engaging in this research is to reveal how understudied physical realities affect humanity's existence. It takes the ordinary and problematizes it to the point of becoming extraordinary, an activity that shows a primary benefit of philosophy's utility. In turn, we gain a view of what philosophical examination can provide, which are insights into how the crisscrossing abstract and concrete subjects yield lessons for both.

While such an undertaking could complicate matters, constructing a "thought model" that examines the interplay of infrastructure, urban sprawl, and NOAs exposes the pattern that lies beneath the conditions that pertain to such situations. Once the issue is situated in this way, we can use the thought model to extrapolate broader principles that expose details about the underlying pattern that pertains to several other cases that involve technology and wicked problems. With such specifics in mind, we see how the conversation about revealing and enframing technologies requires an additional classification. It cannot advance our understanding of how such categories are ill-equipped to see how climate change challenges mental life. Due to this situation, we understand why there is a need for the classification of wicked technologies.

For instance, experts previously thought that southern Nevada did not have naturally-occurring asbestos (Sever 2019). However, local geologists discovered deposits, which alerted managers of an infrastructure project that would meet travelers' and residents' transportation demands, the Boulder City Bypass (Sever 2019). This highway extension would allow tourists to reach nearby Las Vegas without driving through Boulder City, a situation that had created terrible traffic conditions for residents (Saylor 2017). The addition of this highway stretch was predicted to lead to the development of new neighborhoods and additional businesses (Scholenmann 2018).

While initially surveying the issue, it appears benign. However, the danger is that there is no known safe level of asbestos exposure, and there are not any readily available treatments of cures for deadly diseases such as mesothelioma. From diagnosis, life expectancy is typically one year or less (Moore et al. 2008). Other conditions such as asbestosis, lung cancer, and autoimmune issues are possible along with this illness. When it comes to diagnosing diseases such as mesothelioma, one significant complication is that it has incredibly long latency periods, ranging from ten to seventy years from exposure (Frost 2013; Lanphear & Buncher 1992). This reality means that victims will often not know that they have been harmed until decades later.

When it comes to the professions that have seen the highest number of sufferers from asbestos-related illnesses, typically asbestos miners fall victim, along with workers in industries such as shipbuilders and commercial occupations such as insulation installers (Moore et al. 2008). While one can argue that NOAs pose less of a threat due to a reduced concentration of asbestos that one finds in rocks compared to refined sources, there is still no known level of exposure that health officials deem safe (American Cancer Society 2019). This point suggests that *all* exposures can be unsafe, even though most victims had lengthy exposures to industrial sources. In turn, engaging in any kind of activity wherein people can inhale asbestos fibers puts them at risk, and the construction project in Boulder City is no exception. Due to the inherent nature of building infrastructure in such conditions, we must consider that they qualify as characteristics that accompany such technologies. These conditions require that we must add to the terminology that Heidegger provides. The goal is to develop a conception that accounts for these conditions. Examining the case above helps us carry out this task.

Consider, for instance, that infrastructures, such as the one in question, expose something about the concept of enframing. That is to say, the delayed danger that accompanies its construction provides a new element that is not significantly present when Heidegger wrote on revealing and enframing technologies. Hans Jonas (1984) deals with this point in the context of ethical obligations and the requirements for action, but he did not explicitly address its effects on being. This condition requires us to analyze this situation, showing another dimension to today's enframing technologies. It shows that they qualify as another kind of technology that deserves a place in the taxonomy. This point does not suggest that they are not enframing technologies, but they are a kind of enframing technology that features an additional characteristic, "wicked conditions." It accompanies enframing in a context that was relatively unknown when Heidegger was writing on the topic. In the same manner that Jonas (1984) argues that ethical theory cannot address the implied consideration for time when it comes to modern technology and its accumulating effects, this notion also applies to technology's ontological status. This element is the wicked condition that partly defines wicked technologies.

Due to the nexus of infrastructure, urban sprawl, and NOAs outside Las Vegas, we are presented with the notion that the workers who were building it encountered a situation that could have severely affected their health in due time. The issue is not only that we are dealing with enframing technology per se, but we are contending with one that has an extra element, the wicked condition. For the Boulder City bypass, this element was the possible delayed harm that could kill workers several decades after building the enframing technology (the bypass). Although occupational safety regulations in the US require management to respond to the discovery of NOAs, supervisors nevertheless took several measures (Akers 2017). By implementing precautionary protocols that are believed to have reduced exposure levels, hopefully keeping workers safe, management removed or significantly reduced this wicked condition. Reports also indicate that the administration departed from the view that construction must continue for its own sake, arguably taking steps to care for workers' well-being.

According to Dreyfus' interpretation of Heidegger, this thinking could suggest that management formed a 'free relationship' with technology (1997). For instance, they did not pursue it with uncompromising enthusiasm for its own sake, a characteristic associated with an unfree relationship. Instead, they were able to examine several possibilities, choosing to continue construction in a manner that showed some respect and concern for workers and residents' safety. This point does not suggest that they must avoid enframing in all such cases. They can choose to engage in ways that show that they are not limited entirely to courses of action wherein enframing holds steady in an unwavering sense. Such actions exhibit that when dealing with technology while facing harm, there are steps that, if taken, can make the technology less problematic. If we examine the pattern of thinking behind the actions above, then we can apply it to how we think about technology and being in the face of climate change. That is, considering that the accumulating effects of most enframing technologies have helped produce a wicked problem, such technologies should bear this designation.

Consider, for instance, that globally we appropriate seventy-five percent of the world's natural materials to meet cities' needs (Hodson et al. 2012). Infrastructures are a primary conduit for their distribution and management. This notion implies that they are now inherently wicked technologies because they presently have the extra element that exacerbates climate change. While at one point in history, infrastructures were only enframing due to holding natural resources in standing reserve, today, they have gone beyond that point, worsening global climate change.

However, playing a role in climate change is not the only kind of wicked condition. It is an exemplar that, when studied, reveals the pattern of wicked conditions that remain in play for such technologies. In turn, focusing on wicked conditions might be a necessary exploration when there is a nebulous element at hand. Although in some cases, it could be a superfluous step that adds only a little to our understanding of how wicked technologies affect humankind's affairs. For most instances, classifying a technology as "wicked" as we understand it in its socio-eco-political surrounding could be sufficient.

2.5 Challenges to an Expanding Taxonomy for Wicked Technologies

Making a case for wicked technologies suggests that there is a reliable boundary between them and enframing ones. This notion requires that enframing technologies were not always wicked, and we only recently became aware of this quality. We were ignorant of this aspect, but our ignorance does not dismiss the reality that the entire time that we were using them, they were playing a role in the impending climate change that we are living with today. In turn, the business of "wicked technologies" is nothing more than new, useless academic jargon that does nothing more to advance our understanding of technology's progression and impacts.

This objection is formidable, and we can resolve it in one of two ways, although there are perhaps more. The first is to claim that Heidegger simply neglected to consider this aspect of enframing technologies' delayed effects. If this is the case, then it seems that the claims above dealing with wicked problems such as climate change are merely wicked conditions. This point means that enframing technologies were always wicked. Yet, if this were true, then it would suggest that there is no way to return to a position wherein technologies are no longer a threat to the conditions of being in the sense that being could cease to be—in the worst sense. Such a position would entail that we could never alleviate the conditions associated with anthropogenic climate change. However, it seems plausible that we could create the conditions to drastically reduce, slow, impede, or mitigate the effects of climate change—or develop a way to live with them that entirely changes the conversation.

If one of those outcomes occurs, then it seems possible that we could still have enframing technologies, ones that turn the nonhuman world into standing reserve. Considering that such an outcome could become a reality, no matter the odds, it nevertheless shows that wicked technologies are not absolute. This point exhibits that wicked conditions are an inherent feature with which we must always contend. Labeling them in this manner provides the conceptual grounding that could help us create conditions that can help shape our lives.

This point suggests that while wicked technologies can shape life, they also present the opportunity to transform such conditions in a hopeful manner, one that can rescue us from such an unfortunate actuality. On this view, we can escape the exclusive conditions for wicked enframing, becoming one of "saving." Along the same lines wherein technology can be wicked, it can also go against these conditions by saving us from that scenario. This kind of position embodies the power behind the line that Heidegger (1997, p. 333) leaves us with at the end of *The Question concerning Technology*: "But where danger is, grows[,] [t]he saving power also." In the next chapter, I explore the kinds of technology and the thinking behind them, illustrating what it means to qualify as a saving technology.

2.6 Conclusion

Although the intersections of developing infrastructure and encountering NOAs is an unusual occurrence that counts as an outlier situation, it shows that nonhuman nature will continue to challenge us in unexpected ways. The significance of this notion rests on the reality that the ingenuity that drives us to move forward is equipped to deal with situations that arise that are conditions of successful creativity and resourcefulness. Viewing such cases in tandem, one way to see such conditions is either solving two problems or one issue with interconnected elements on the way to task completion. Either we solve a complex problem or multitask. The point here is that the kind of thinking that guides our technological actions can overcome these hurdles, revealing that understanding difficulties means addressing dimensions that might not arise until the project that will advance our primary objective is underway.

Aside from the choices associated with conundrums' outcomes, thinking that relies on "saving power" can accompany the actions that reflect optimal contemplation. This idea suggests that we cannot only solve problems, but we can also examine a range of possibilities. This aspect is of the utmost importance because any choice displays what the people behind the decision chose as the best from the full range of options, meaning that such reasonings carefully examined steps required for progress. There is no reason not to go with one that aligns theoretically with outcomes that lend themselves to creating opportunities to shape conditions for our mental lives favorably.

The following chapters begin focusing on that enterprise. They show that while moving towards technological progress benefits from an expanding ontological consultancy, the mere idea of bringing the thought patterns behind saving power into our mindset requires conceptualizations that can also move it forward. Otherwise, the term only serves as a static demonstration of pioneering intellect that differs from its principled ambition. If there is a worthwhile goal to advance our understanding of the motivation to drive future endeavors to deliver conditions for mental life that steer away from those that created outcomes that we now know to avoid, pursuing it might not yield immediate benefits. However, it aims to correct the course, which must hold steady if we wish to secure alternative circumstances for the chances of our continued survival. Ultimately, no less than this intense reality is at stake.

References

Akers M (2017) As Interstate 11 progresses, asbestos monitoring continues. Las Vegas Sun May 23. Retrieved from: https://lasvegassun.com/news/2017/may/23/as-interstate-11-progresses-asbestos-monitoring-co/

American Cancer Society (2019) Asbestos and Cancer Risk. https://www.cancer.org/cancer/cancer-causes/asbestos.html

Berry H, Bowen K, Kjellstrom T (2010) Climate change and mental health: a causal pathways framework. Int J Public Health 55(2):123–132

Clayton S (2020) Climate anxiety: psychological responses to climate change. J Anxiety Disord 74:102263

Collingridge D (1982) The social control of technology. St. Martin's Press, New York

Dana D, Tuerkheimer D (2016) After Flint: environmental justice as equal protection. Northwest Univ Law Rev 111(3):879–890

Davies K, Fisher K, Dickson M, Thrush S, Le Heron R (2015) Improving ecosystem service frameworks to address wicked problems. Ecol Soc 20:2

Dreyfus H (1997) Heidegger on gaining a free relation to technology. In: Shrader-Frechette K, Westra L (eds) Technology and values. Rowman & Littlefield Publishers, Lanham, MD, pp 41–54

Epting S (2020) Infrastructure, Urban Sprawl, and naturally-occurring asbestos: an ontological thought model for wicked and saving technologies. Open Philos 3(1):389–399

Frost G (2013) The latency period of mesothelioma among a cohort of British asbestos workers (1978–2005). Br J Cancer 109(7):1965–1973

Heidegger M (1993, 1997) The question concerning technology. In: Krell D (Ed) Martin Heidegger: basic writings. HarperCollins Publishers, New York, pp 311–341

Hodson M, Marvin S, Robinson B, Swilling M (2012) Reshaping urban infrastructure: material flow analysis and transitions analysis in an urban context. J Ind Ecol 16(6):789–800

Irwin R (2010) Climate change and Heidegger's philosophy of science. Essays Philos 11:16–30

Jonas H (1984) The imperative of responsibility. in search of an ethics for the technological age. Chicago University Press, Chicago

Lanphear B, Buncher R (1992) Latent period for malignant mesothelioma of occupational origin. J Occup Med 34(7):718–721

Manning C, Vallyathan V, Mossman B (2002) Diseases caused by asbestos: mechanisms of injury and disease development. Int Immunopharmacol 2.2(2):191–200

McCrea D (2014) Asbestos proves to be a microscopic road block near Boulder City. Las Vegas Sun July 20. https://lasvegassun.com/news/2014/jul/20/asbestos-microscopic-road-block/

Moore A, Parker R, Wiggins J (2008) Malignant Mesothelioma. Orphanet J Rare Dis 3(1):34

Myers T (2014) Understanding climate change as an existential threat: confronting climate denial as a challenge to climate ethics. De Ethica. J Philos Theol Appl Ethics 1(1):53–70

Nagenborg M (2018) Urban robotics and responsible urban innovation. Ethics Inf Technol 22:345–355

Payne B (2015) Oral history of Bonton and ideal neighborhoods in Dallas, Texas. MA Thesis, University of North Texas

Perkins R, Hargesheimer J, Vaara L (2008) Evaluation of public and worker exposure due to naturally occurring asbestos in gravel discovered during a road construction project. J Occup Environ Hyg 5(9):609–616

Peters B (2017) What is so wicked about wicked problems? A conceptual analysis and a research program. Policy Soc 36(3):385–396

Rittel H, Webber M (1973) Dilemmas in a general theory of planning. Policy Sci 4(2):155–169

Saylor H (2017) Bypass blends with environment. Boulder City Review May 24. https://boulderci tyreview.com/news/bypass-blends-with-environment/

Seto K, Güneralp B, Hutyra L (2012) Global forecasts of urban expansion to 2030 and direct impacts on biodiversity and carbon pools. Proc Natl Acad Sci 109(40):16083–16088

Sever M (2019) Asbestos fibers thread through rocks and dust outside Vegas. Earth and Space Science News November 6. https://eos.org/articles/asbestos-fibers-thread-through-rocks-and-dust-outside-vegas

Schoenmann J (2018) On verge of a boom, boulder City Keenly aware of asbestos, Nevada Public Radio. https://knpr.org/knpr/2018-10/verge-boom-boulder-city-keenly-aware-asbestos

Scott A (2012) By the grace of god. Portland Monthly, March. https://www.pdxmonthly.com/news-and-city-life/2012/02/african-american-churches-north-portland-march-2012

Srinivasan S, O'fallon L, Dearry A (2003) Creating healthy communities, healthy homes, healthy people: initiating a research agenda on the built environment and public health. Am J Public Health 93(9) 1446–1450

Winner L (1980) Do artifacts have politics? Daedalus 109(1):121–136

Chapter 3
Cities and Mental Life

Abstract While climate change can induce mental states such as anxiety, fear, stress, and depression, which affect our existence, other wicked problems such as vast social and economic inequalities that manifest in the urban environments do also. This phenomenon appears to be well entrenched in many cities today. In turn, through expanding our taxonomy to include wicked and saving technologies, we can employ these terms to enhance our understanding of situations in the urban sphere that manifest as distinctly harmful mental states. Common examples include homelessness and home-related issues, traffic, and extreme loneliness. Due to passive acceptance of such situations, some people might assume that these unfortunate realities are metropolitan places' inherent characteristics. However, viewing them as merely circumstantial provides a way to change how we perceive cities' possibilities, meaning that we can upgrade our position on such matters. This chapter lays the groundwork for that undertaking. It identifies some of the wicked technologies in play, and in turn, marks them as ones that could benefit from specialized study to discover means that can mitigate harms and change the unwelcome conditions surrounding them.

Keywords Mental states · Social inequity · Mental life · Zoning · Transportation · Housing

3.1 Introduction

Having established familiarity with wicked technologies in the previous chapters, showing how climate change creates additional elements that require attention, this concern is not the only one that needs advanced study. There is a lengthy history of vast social and economic inequalities in cities where such conditions express themselves in distinct forms that induce harmful mental states such as anxiety, stress, fear, and anger. One problem is that their familiarity could mean that people accept them as if they were urban characteristics that one usually associates with metropolitan places. For instance, seeing homeless people in global cities is customary: nothing to see here—move along.

© The Author(s), under exclusive license to Springer Nature Switzerland AG 2021
S. Epting, *Saving Cities*, SpringerBriefs in Philosophy,
https://doi.org/10.1007/978-3-030-85833-9_3

The problem is that mere familiarity does not entail that such conditions are permanent or should be that way. This chapter begins the process of identifying other manifestations of wicked problems and troubling outcomes that bring unwanted mental states such as those mentioned above. To make this case, it turns back over a century, examining George Simmel's thoughts on how the metropolis delivers fewer promising avenues for human interactions. Then, the investigation moves forward, looking at a more recent testimony that expresses urban frustrations and clinical studies that show harmful experiences.

Having grounded these issues by illustrating their entrenchment in contemporary cities, I examine a few of the troubling technologies that play a more prominent role in wicked problems of concern: zoning, housing, and transportation. Although these affairs are not the only ones that matter, they play significant roles in shaping mental life. After identifying these situations and studying their effects, the stage is set for the following chapter that examines the kind of thinking that can foster technological arrangements in urban settings that favor more desired mental states.

3.2   Wicked Problems and Mental Life

While one can fathom several reasons why cities can induce widespread anxiety that rests on social and economic inequalities, such situations are wicked problems (Rittel & Webber 1973). Recalling from Chap. 2, this point suggests that we cannot pinpoint their precise origins, the matter is worsening, humankind is responsible, and the pathway out of it remains unclear, among other aspects. Philosophers and interdisciplinary researchers can argue about grand notions such as capitalism and its role in such matters, but it seems challenging to argue with people's lived experiences. Even if we cannot reach a consensus on the origins or solution to such problems, there is no good reason to deny people's testimonies about how cities shape their proclivities towards the mental states that arise from socio-material arrangements of structures and infrastructures. Although this issue holds significant importance for contemporary society, it has an extensive history, which shows that this issue tends to accompany urbanity.

Consider, for example, Simmel's (1903) *The Metropolis and Mental Life*. Throughout the text, Simmel details how the built environment creates conditions wherein the city is a burden that residents must shoulder—an "antisocial" social structure. Compared to the countryside, the city remains inherently antagonistic towards the concept of community. The only option for social interaction is relating to each other and the city through money (Simmel 1903). Simmel (1903, p. 12) holds:

> The modern city, however, is supplied almost exclusively by production for the market, that is, for entirely unknown purchasers who never appear in the actual field of vision of the producers themselves. Thereby, the interests of each party acquire a relentless matter-of-factness, and its rationally calculated economic egoism need not fear any divergence from its set path because of the imponderability of personal relationships. This is all the more the case in the money economy which dominates the metropolis in which the last remnants

of domestic production and direct barter of goods have been eradicated and in which the amount of production on direct personal order is reduced daily.

The passage above reveals that when the market drives urban complexity, it transforms humanity into something that works against itself, an urban system that exists solely for the economy's benefit. Due to this condition, all other elements of city life become secondary considerations. This notion suggests that essential concerns such as human health, justice, wellbeing, and flourishing must receive consideration after economic forces. One could argue that urban dwellers in the above case exist as standing reserve for the capitalist city, making little sense when the city should serve the people. The outcome, then, is the blasé attitude that such cities can produce, as was Simmel's take on life in the metropolis (Simmel 1903). Due to how cities can overwork the senses, city folk leans toward indifference to urban existence's facets. This point shows that familiarity, even when it involves grasping changing stimuli constantly, can yield a kind of dullness, a quality that does not warrant pursuit.

While the point here is not to reduce the analysis solely to a critique of capitalism, one can say that viewing a city's purpose as only to serve the market would have extremely adverse effects. Again, the point here is not to merely identify the causes of these conditions because they will vary from one location to the next, but the situations that shape mental states are of immediate interest. This notion does not hold that the roots of the problem do not deserve attention. The focus remains centered on how they favor mental states such as anxiety. Central to this idea is that many urban dwellers' mental lives include anxiety, fear, dread, and similar mental states. Simmel (1903, pp. 11–12) describes this outcome in the following way:

> To the extent that the metropolis creates these psychological conditions—with every crossing of the street, with the tempo and multiplicity of economic, occupational and social life—it creates in the sensory foundations of mental life, and in the degree of awareness necessitated by our organization as creatures dependent on differences, a deep contrast with the slower, more habitual, more smoothly flowing rhythm of the sensory- mental phase of small town and rural existence. Thereby the essentially intellectualistic character of the mental life of the metropolis becomes intelligible as over against that of the small town which rests more on feelings and emotional relationships.

Examining contemporary urban life through a Simmel-inspired lens, we see that traffic jams, air pollution, crime, and overcrowding might lead to the conditions that make metropolitan life harmful to people's minds. These conditions predispose them to mental states such as anxiety that make life miserable. Such arrangements can hurt people in numerous ways, and the longevity of such harms can become so deeply entrenched in urbanity so that they are passed on from one generation to the next. In turn, along with genetic characteristics, people inherit their cities' socio-political conditions and the preconditions for anxiety that the city yields.

Although Simmel's insights are over one hundred years old, they remain just as relevant today for many urban environments, subject to the changing conditions particular to any given setting. Despite writing the above passage with the influence of a specific metropolis in mind, the pattern behind its tendency to create such conditions for mental life is widely applicable across cultures, religions, economies, and global positions. For a more modern testimony, 1980's hip-hop pioneers Grandmaster Flash and the Furious Five illustrate the severity of the urban condition in their classic hit, "The Message" (Chase et al. 1982). This song illustrates urban scenarios that are highly unpleasant and foul-smelling, including having to deal with unfavored insects and pests, drug addiction, economic hardship, and other matters (Chase et al. 1982). The chorus pleads for those near the person not to provoke him because he is attempting to thwart a breakdown in his cognitive functions (*ibid.*).

Although the song's description is an artistic expression of the urban condition (albeit anecdotal), studies suggest that there is a psychological basis underpinning such views. Granted, they are referencing a specific city, meaning that other urban environments have unique conditions. Still, it seems reasonable to hold that similar situations will produce the same kind of effects on mental life. Researchers who work across the academy investigate the impacts that different parameters of urban living can have on residents' mental health (e.g., Lewis & Booth 1994; Lambert et al. 2015; Gold et al., 2019).

Although this point has sometimes drawn criticism, an abridged review of the literature suggests that we should not abandon studies that focus on the connection between cities and mental illness. For instance, Glyn Lewis and Margaret Booth (1994) discover an association between urban dwelling and psychiatric morbidity. Lambert et al. (2015) hold that some urban dwellers might be at risk for developing anxiety issues and mood disorders, and they also exhibit how urban problems such as pockets of extreme poverty and urban design could raise concerns for mental illness. Additionally, Gold et al. (2019) are currently investigating the association between living in the city and mental illnesses such as schizophrenia and psychosis.[1]

As mentioned earlier, taking inventory of the characteristics associated with harmful mental states does not require a particular identifier because the outcomes illustrate that a problem exists. There is a correlative response that gives way to alternative realities. Such notions speak to people's everyday lived-world fears of factual matters. Several disciplinary studies confirm the significance of these affairs.

For instance, cities with food insecurity problems have received numerous studies on how such conditions increase anxiety, particularly in marginalized groups who are already enduring compounding hardships. This subject brings multiple concerns into view: health, fitness, pediatric nutrition, and wellbeing, to name a few. The point here is not that people lack food, but their present conditions are such that access to healthy options remains challenging, and permanent solutions are not often on the horizon. Consider, for instance, that Jane Battersby and McLachlan (2013, p. 716) explain how urban design perpetuates ingrained, systematic food problem with far-reaching effects for residents:

---

[1] It is worth mentioning that despite the indications that urban residence could be linked to increased risk of mental illness, research from Breslau et al. (2014) suggest that there is little difference between the ways that urban and rural environments affect people's mental lives, a new matter that speaks to Simmel's view over one hundred years ago. However, this point does not suggest that urban places do not affect people's mental lives, which suggests that they require specific mitigatory efforts to target their ill-effects appropriately.

SA [South African] cities are shaped by their colonial and apartheid histories. By design the lowest income households continue to be located on the peripheries of the cities. This urban geography influences food insecurity in several ways. Firstly, it shapes economic opportunities and the potential to earn income for residents of low-income areas. In a market economy, income is a major driver of food security. Low-income earners, living long distances from jobs and facing inadequate public transport, have lengthy commutes and spend a large proportion of their incomes on transport, leaving less money to buy food. With limited time, and rising energy costs, households choose to cook foods that require less preparation and use more processed and prepared foods. These foods are often more expensive per unit and less nutritionally dense than more traditional foods. Food choice is therefore not simply a matter of personal choice, but also of urban design.

The passage above illustrates that food security can be a deeply entrenched wicked problem according to established criteria as examined in the previous chapter. Even though we can isolate such affairs as problems of X or Y, they remain entangled within larger wicked problems, resting on a particular cities' historical conditions. Although food security is one pressing matter, the pattern underlying it can extend into any urban affair, revealing that while issues in a metropolitan environment can benefit from examination in isolation, completely doing so without consideration for neighboring elements could be wrongheaded.

In addition to food matters, conversations on homes and housing raise several concerns. Numerous cities face challenges of homelessness wherein vulnerable populations live with conditions such as chronic stress (Klitzing 2003; Kirkman et al. 2010). Displacement due to gentrification and similar situations is another source of stress (Smith et al. 2020). When it comes to renting increases for apartment dwellers, keeping up with soaring costs while balancing other needs can also increase worry, especially for seniors (Khouri & Shalby 2019).

While these issues are apparent to the people experiencing them and are frequent topics of urban interest, other problems that require a bit more understanding of urban planning also play a significant role in how cities have developed, especially regarding zoning. For example, single-family zoning has primarily facilitated the process of keeping cities, suburbs, and towns restricted to certain forms. The kind of thinking behind such measures has significant impacts on housing options available in an area, relegating multi-family dwellings to areas that are usually separated by distance. On the surface, these situations are benign. However, such zoning requirements count as substantial factors in larger wicked problems (Serkin 2020). In turn, there has been a significant push in the urban planning literature to end single-family zoning (Manville et al. 2020).

For instance, Manville et al. (2020, p. 106) zero in on the reasons why researchers and professionals are calling for bold efforts:

Most American cities have a zoning designation prohibiting all development except detached single-family homes. Many cities apply this designation to most of their land. We think this designation, usually called R1, should not exist. R1 is inequitable, inefficient, and environmentally unsustainable. It lets a small number of people amass disproportionate property wealth, excludes many others from high-opportunity neighborhoods, and forces others to pay more for housing than they should (Lens & Monkkonen 2016; Reeves 2017). R1 was born from, and codifies, base and tribal instincts: a desire to set privileged in-groups apart and keep feared or despised out- groups at bay (Nightingale 2012). Its history is explicitly classist and deeply interwoven with racism, and its present form only barely conceals these origins (Rothstein 2017; Trounstine 2018; Weiss 1987). It should have no future. Planners should actively work to end it.

Regarding the impacts on people struggling to live in locations where single-family zoning is an overly burdensome issue, modest apartments or similar arrangements are systematically excluded (Manville et al. 2020). In turn, the economic pressures cause debt-related stress and anxiety for families aiming to secure financial stability (*ibid.*). For such cases, zoning is a wicked technology to the extreme, and future studies remain paramount to deliver mitigatory efforts. These families need saving technologies.

While some troubling aspects associated with wicked housing are well-known, others are hidden from view entirely, meaning that people suffer, and it goes mostly unnoticed. One such instance is an emerging problem for mental states is perceived social isolationism, wherein people feel incredibly lonely—even though they remain surrounded by people essentially (Holt-Lunstad et al. 2015). This issue is of particular interest for advanced study because it could easily blend into the cityscape. People's constant exposure to such conditions might normalize it to the extent that alternatives seem plain wrong.

For an issue with a contrasting scale, car-centric cities are a monumental problem, creating numerous matters that intersect with physical and mental health in the city (Shaver 2016). Yet, innumerable cities' configurations force people to deal with these affairs daily. Some scholars refer to these circumstances as instances of forced-car ownership (Currie & Senbergs 2007; Miller 2009). This research shows that people have no option but to deal with the troubles mentioned above as an exclusive element that comes with navigating the urban sphere. In turn, the mental states associated with automobiles are entrenched into mobility. Subjects such as "road rage," which affects numerous people, remain inextricably linked to car-centric places (James 2009).

What is more, one can argue that by giving so much attention to personal vehicles, public transit often is under-utilized (Giuliano 2005). Such situations force people to deal with grossly inadequate services that produce states of worry and stress. Studies now illustrate how the noise from personal vehicles is harmful (McAlexander et al. 2015). Roadways and gas guzzlers are the primary means of traversing metropolitan environments, meaning that they play a significant role in determining urban aesthetics (Wright & Curtis 2020). For instance, some urban places have horrendous "spaghetti junctions" defined as expansive highway exchanges that produce the view of massive pasta noodles from high above (Moran 2010). Not only is such a scenario unattractive, one could argue, but all other transportation-related and neighboring aesthetics must work around their presence. This point entails that efforts to beautify cities must confront these situations.

While the above list of urban issues remains non-exhaustive, it indicates the sort of problems that one could expect to emerge that play a role in inducing harmful mental states such as anxiety, stress, anger, and dread. The notion to bear in mind is that people are often dealing with these issues simultaneously—along with others not mentioned. This point shows that despite the lack of attention to ethical ideas in this work, they are always there, even though they require separate investigations to do them justice.

Still, aside from claims about cities and their possible impacts on mental health, the view that cities can contribute to stress seems straightforward, and researchers are searching for ways to design them so that they burden the mind to a lesser degree (Bakolis et al. 2018). For traditional cities, one could argue that re-creating them to accomplish the above task would be controversial, expensive, and perhaps highly challenging for urban planners.

Despite this sobering reality, the challenge is one that urban professionals and residents should look at with enthusiasm. It can create mental states that deter harmful ones, as mentioned earlier. To show how such situations can arise, a view illustrating how the city can serve as the center for activities that affect our lives in myriad ways must come into focus. Enhanced positions that highlight this signature quality of cities make this point evident. In turn, the following section does just that. It brings insights from Lewis Mumford into the examination, showing how cities can facilitate several social exchanges, determining our mental lives' contours.

3.3 The City as Stage for Mental States

It makes sense to say that the city can provide any number of experiences and feelings, putting us in touch with mental states. The ideas and situations previously explored such notions in terms of wicked problems and harmful urban arrangements. However, for those events to take place, they need to rest on a larger stage that makes them possible, which, for the case at hand, is the city. Their primary structures help our lives take shape. For example, Lewis Mumford (1937, p. 59) illustrates the significant role that cities play in all such affairs:

> The city in its complete sense, then, is a geographic plexus, an economic organization, an institutional process, a theater of social action, and an aesthetic symbol of collective unity. The city fosters art and is art; the city creates the theater and is the theater. It is in the city, the city as theater, that man's more purposive activities are focused, and work out, through conflicting and cooperating personalities, events, groups, into more significant culminations. Without the social drama that comes into existence through the focusing and intensification of group activity there is not a single function performed in the city that could not be performed—and has not in fact been performed—in the open country. The physical organization of the city may deflate this drama or make it frustrate; or it may, through the deliberate efforts of art, politics, and education, make the drama more richly significant, as a stage-set, well-designed, intensifies and underlines the gestures of the actors and the action of the play.

From this passage, we see what the city can become and why it holds such significance. Bearing in mind that most of the world's people now inhabit cities, and researchers predict this trend will continue, each city serves as a chance to improve the conditions associated with residents' mental lives. The city, as a technology, can save us from only knowing and thinking about people and nonhumans as standing reserves without reservation.

As mentioned, this situation is what we need saving from, and it underscores why such pursuits must begin with an orientation that inherently goes against it by design. This kind of thinking is not merely an attitude. We should see it as a saving technology itself, considering that it is an invented way for reasoning that aims for a particular product—an outcome wherein we create cities that favor mental states that focus on the positive while diminishing the adverse effects. In turn, such thinking can help usher in better cities that improve the urban condition.

3.4 Conclusion

This chapter focused on the needed mindset to deliver the required outcomes. The following chapter begins with the idea that ways of thinking, as touched on in this chapter, are also technologies—*designed* cognitive "devices" that help achieve task completion. These mental apparatus help secure the conditions for saving cities, even though they might have situational shortcomings. In turn, their purposes are twofold. They account for the excellent mindset geared towards transforming the urban sphere, and they are saving technologies at the same time. This notion holds that a saving city is one that, in a loose sense, reflects such thinking through its configuration.

References

Bakolis I, Hammoud R, Smythe M, Gibbons J, Davidson N, Tognin S, Mechelli A (2018) Urban mind: using smartphone technologies to investigate the impact of nature on mental wellbeing in real time. Bioscience 68(2):134–145

Battersby J, McLachlan M (2013) Urban food insecurity: a neglected public health challenge. SAMJ: South African Med J 103(10):716–717

Breslau J, Marshall G, Pincus H, Brown R (2014) Are mental disorders more common in urban than rural areas of the United States? J Psychiatr Res 56:50–55

Chase C, Fletcher E, Glover M, Robinson S (1982) The message. Recorded by grandmaster flash and the furious five. On The Message, Sugar Hill, Englewood, New Jersey

Currie G, Senbergs Z (2007) Exploring forced car ownership in metropolitan Melbourne. Australasian Trans Res Forum 2007

Giuliano G (2005) Low income, public transit, and mobility. Transp Res Rec 1927(1):63–70

Gold I, Bornstein L, Choudrey S, Goméz-Carillo A, Shah J, Weinstock D (2019) Cities and the mind. McGill Inst Health Social Policy. Available online. https://www.mcgill.ca/ihsp/research/cities-and-mind

Holt-Lunstad J, Smith T, Baker M, Harris T, Stephenson D (2015) Loneliness and social isolation as risk factors for mortality: a meta-analytic review. Perspect Psychol Sci 10(2):227–237

James L (2009) Road rage and aggressive driving: steering clear of highway warfare. Prometheus Books, Amherst, NY

Khouri A, Shalby C (2019) Seniors facing eviction fear homelessness and isolation as California's housing crisis rolls on. Los Angeles Times, August 29. https://www.latimes.com/business/story/2019-08-28/senior-housing-crisis-impact

Kirkman M, Keys D, Bodzak D, Turner A (2010) Are we moving again this week? Children's experiences of homelessness in Victoria Australia. Social Sci Med 70(7):994–1001

Klitzing S (2003) Coping with chronic stress: leisure and women who are homeless. Leis Sci 25(2–3):163–181

Lambert K, Nelson R, Jovanovic T, Cerdá M (2015) Brains in the city: neurobiological effects of urbanization. Neurosci Biobehav Rev 58:107–122

Lens M, Monkkonen P (2016) Do strict land use regulations make metropolitan areas more segregated by income? J Am Plann Assoc 82(1):6–21

Lewis G, Booth M (1994) Are cities bad for your mental health? Psychol Med 24(4):913–915

Manville M, Monkkonen P, Lens M (2020) It's time to end single-family zoning. J Am Plann Assoc 86(1):106–112

McAlexander T, Gershon R, Neitzel R (2015) Street-level noise in an urban setting: assessment and contribution to personal exposure. Environ Health 14(1):1–10

Miller M (2009) Forced car ownership and transport disadvantage: a spatial analysis of transport and equity in a regional city, Dissertation, Monash University

Moran J (2010) On roads: a hidden history. Profile Books, London

Mumford L (1937) What is a city? Archit Rec 82(5):59–62

Nightingale C (2012) Segregation: a global history of divided cities. University of Chicago Press, Chicago

Reeves R (2017) Dream hoarders. Brookings Institution, Washington, DC

Rittel H, Webber M (1973) Dilemmas in a general theory of planning. Policy Sci 4(2):155–169

Rothstein R (2017) The color of law. Liveright (Norton), New York, NY.

Serkin C (2020) The Wicked Problem of Zoning. Vanderbilt Law Rev 73(6)

Shaver K (2016) Suburbs increasingly view their auto-centric sprawl as a health hazard. *Washington Post*, December 28. https://www.washingtonpost.com/local/trafficandcommuting/suburbs-increasingly-view-their-auto-centric-sprawl-as-a-health-hazard/2016/12/28/49a99542-c6f9-11e6-8bee-54e800ef2a63_story.html

Simmel G (1903, 2002) The metropolis and mental life. In: G. Bridge and S. Watson (eds) The Blackwell City Reader, Wiley-Blackwell, Oxford and Malden, MA

Smith G, Breakstone H, Dean L, Thorpe R (2020) Impacts of Gentrification on Health in the US: a Systematic Review of the Literature. J Urban Health 97(6):845–856

Trounstine J (2018) Segregation by design: local politics and inequality in American cities. Cambridge University Press, Cambridge, UK

Weiss, M. (1987) The rise of the community builders: The American real estate industry and urban land planning. Columbia University Press, New York, NY

Wright C, Curtis B (2020) Aesthetics and the urban road environment. In: Proceedings of the institution of civil engineers-municipal engineer, vol 151, issue no 2, pp 145–150

Chapter 4
Saving Technologies

Abstract While the previous chapter began expanding the taxonomy of technologies so that their impacts on mental life surfaced, this one continues that process. It shows that to move forward, we need a conceptual model that identifies kinds of technology dissimilar from previous ones and qualifies as having characteristics that can deliver desired outcomes. The present case includes the mitigatory efforts and innovative spirit that speak to the pattern behind saving power. This chapter does just that. It illustrates that such progress, considering wicked problems and impossible conditions, requires thinking about these affairs in this fashion. Even though such exploration reveals that saving technologies could serve us well in our pursuits, the modern temper will find them unfamiliar. However, if the goal is to save us from ourselves, we must understand the reasons for their embrace, which is what this chapter does, culminating in a push to move from saving technologies to saving cities.

Keywords Saving power · Taxonomy · Saving technologies · Saving cities

4.1 Introduction

Having established an understanding of the complexity behind technology's taxonomy, moving forward will show that the future should not resemble the past. This point does not suggest that we cannot make use of the lessons that it provides. Still, the attention must remain centered on developing the mindset to secure better conditions for enhancing mental life. This notion entails unpacking such an account, which is the aim of this chapter.

After seeing the illustrations that such an investigation provides, the next step is to survey the kind of thinking that indicates shifting away from the thought patterns associated with enframing technologies to ones that can save us from ourselves, "saving technologies." Following this study, I address some concerns that this exercise raises. Next, I introduce the idea that saving technologies can lead to saving cities, a move the prepares us to fully address this notion in the chapter that follows this one.

S. Epting, *Saving Cities*, SpringerBriefs in Philosophy,
https://doi.org/10.1007/978-3-030-85833-9_4

4.2   Towards Saving Technologies

Advancing our thinking means that we must also learn to live in a manner that demands rigorous honesty about which technologies challenge or support ways of thinking that reject or embrace enframing or wicked technologies. We must now develop technologies that can save us (literally), reflecting better ways of thinking. Dealing with saving technologies requires us to change our lens to gain a forward-looking perspective. This orientation does not dismiss the harms that stem from the past, but they do not overly dwell on aspects such as assigning blame.

Instead, the primary focus is to identify a way forward—with our thinking, which alters the conditions for mental life to discontinue the kind of thinking that yields wicked technologies. Such a move would create "better" conditions for mental life. While this quality and its prioritization are not essentially required for mitigating the harms associated with environmental degradation or social injustice, it is one pathway to changing the conditions that influence mental life. It respects Heidegger's warning, creating an alternative to harmful monological thinking. Shifting to such a mindset, however, could help in alleviating the problems above.

The thought model that we can derive from the actions in southern Nevada following the discovery of naturally occurring asbestos does just that. It provides an opportunity to see how we can use lessons from this case to learn about applying extrapolated insights to improve the conditions for influencing mental life, which takes Heidegger's warning seriously. The motivation behind employing saving technologies as part of an expanded taxonomy of technologies is to categorize technology to provide the means for reshaping the conditions for maintaining our existence that remain unrestricted. For instance, Dreyfus (1997, pp. 50–51) holds that, for Heidegger, forming a free relationship with technology requires that we can embrace practices of the past that are not associated with the thinking that governs enframing exclusively:

> Heidegger would say that we should, indeed, try to preserve such practices, but they can save us only if they are radically transformed and integrated into a new understanding of reality. In addition, we must learn to appreciate marginal practices—what Heidegger calls the saving power of insignificant things—practices such as friendship, back-packing into the wilderness, and drinking the local wine with friends. All these practices are marginal precisely because they are not efficient. They can, of course, be engaged in for the sake of health and greater efficiency. This expanding of technological efficiency is the greatest danger. But these saving practices could come together in a new cultural paradigm that held up to us a new way of doing things, thereby focusing a world in which formerly marginal practices were central and efficiency marginal.

This passage holds importance for at least two reasons. Firstly, it shows that charting a pathway forward can save us from an ill fate. We can select devices, resources, and procedures that give defining features to individuals' worlds and, in light of today's circumstances, remain premised on the reality of dealing with a wicked and uncompromising problem, climate change. This point does not entail that we can employ enframing technologies with impunity. However, the notion that deserves underscoring here is that we can choose the option based on its ability to secure saving conditions. Due to this consideration, we must employ a free relationship to develop the practices which will save us from an anthropogenic demise.

This idea suggests that, for cases in Nevada, for instance, we can choose to engage with arrangements of technologies that yield favorable or at least preferable outcomes—given the circumstances. For the case in Nevada, completing the infrastructure project provided traffic relief to nearby residents, which could decrease emissions from idling vehicles. The discovery of the NOAs happened while the project was well underway. The steps taken suggest that thoughtful deliberation was present, gesturing towards a free relationship, one that favored a highway bypass to mitigate some harm. The takeaway here is that while saving technologies could be optimal, there are instances wherein we should strive for progress despite knowing that we might not achieve perfection.[1] Still, such efforts bolster the global condition of collectively contending with the effects of climate change and perhaps other wicked problems.

Secondly, while the saving power of insignificant things was applicable before the wicked problem of climate change, bolder efforts are required now. We need the saving power of significant things to reduce the threats that give us hope, which, one could argue, serves as the antithesis to anxiety. Infrastructure can fit in this category due to the monumental impact that it has on transforming our lives. Urban infrastructures include physical technologies such as transport systems, recycling facilities, and sanitation systems. They also include digital components such as information-communication technologies (ICT) that help municipalities monitor and control a city's services. These devices would generally qualify as enframing technologies.

However, if the thinking behind them exhibits concern for reducing threats, one that provides hope, and we can verify positive impacts, then the label "saving" is appropriate. However, it is imperative that, when applying this term, it is honest. This process requires significant evaluation because it concerns the basis of a free relationship with technology. One worry is that while moving away from seeing entities as standing reserve that aligns with wicked technology, pursuing an option that involves standing reserve to eliminate harmful conditions is bad or worse. In turn, it appears as if we have embraced standing reserve as part of a free relationship to avoid thinking in such a way, but it is merely another version of wicked thinking.

Aside from this worry, numerous approaches embody the kind of mindset that can help form a free relationship, revealing the necessary attitude to maximizing its saving power. Analyzing such approaches to determine if they possess this quality makes good use of philosophy. However, scientists, engineers, planners, other researchers, and professionals must evaluate the outcomes or their likeliness to ascertain if a device has the physical properties and the social situatedness that would qualify it as a saving technology or not. This point also shows that we must deal with

---

[1] It is worth mentioning that some researchers in the neighboring area of value-sensitive design also favor the view of "progress-not-perfection." For more information, see Friedman and Hendry (2019). *Value sensitive design: Shaping technology with moral imagination.* Cambridge, MA: MIT Press.

tame elements in a wicked world. The wicked are always there, lurking over the problem almost as if it there to remind us that our best scientific thinking demands the company of philosophical analysis to employ such lessons effectively. Otherwise, we are stuck with knowledge without know-how—a failure to launch the culmination of the world's expertise.

Similar to how learned minds discover new scientific knowledge, philosophically based enterprises continue to advance the kind of thought required to deal with the wicked. It addresses how to conduct, shape, pursue, think about, and direct the lessons that the sciences and engineering provide. While they also must demonstrate their abilities to deal with a wicked world's conditions, they favor the kind of thinking designed for the wicked. In turn, the governing pattern that gives them structure helps form a free relationship with technology, meaning that we can better determine how we employ technology for our sake rather than for its own. When embracing such an orientation, we increase the number of possibilities for thinking about complex relationships with and through technology. This point means that identifying how urban technologies help produce mental states that play a role in shaping the quality of our mental lives in the face of wicked problems such as climate change.

While there are numerous ways to identify the early foundations of such thinking (e.g., Aldo Leopold's land ethics. See: Leopold 1949, 1968), several methods illustrate its defining features. They seek to discover ways to interact with the nonhuman and human worlds that exhibit careful deliberation. For example, sustainable urbanism, bio-urbanism, cradle-to-cradle design, and urban resiliency have attitudes that reflect a mindset that suggests that their design's motivation is not entirely consistent with enframing.[2] Even though they differ when comparing their internal structures, the thinking behind them endeavors to mitigate environmental and social harm.[3] This essential quality goes against the inherent characteristic that we associate with enframing, as Dreyfus (1997) points out. When we examine many of the approaches mentioned above, we find numerous examples of devices that align with the attitude required for counting as saving technologies.

For instance, urban infrastructures such as mass transit, safe bicycle lanes, Earthships, and cohousing initiatives reflect the kind that ecologically hopeful thinking that can mitigate environmental harm. Although "cohousing communities" can count as saving technologies, the term "cohousing communities" is also a saving technology, a saving technology of thought. This notion means that it has the characteristics that play a role in mitigating particular wicked problems. One could dismiss this notion as being insignificant, but such a position would be misguided. Saving thought technologies serve as the foundation that secures the building blocks' positioning that will save us. Each cohousing community, when situated properly, counts as one such

---

[2] For an example of sustainable urbanism, see Farr (2011). For an example of Bio-urbanism, see.
Tracada and Caperna (2013). For an explanation of cradle to cradle, see McDonough and Braungart (2010). For a review of urban resilience, see Meerow (2016).

[3] It is worth mentioning that, according to McDonough and Braungart (2010) holding the view that making industry less bad is inherently flawed and limited because it does not support needed environmental change.

block. The point is to replace, rebuild, and or build on sturdy ground. Starting with saving thoughts does just that.

While the thinking behind these saving technologies provides an alternative to the sort of thought associated with enframing, there is more to them than is immediately evident. The saving power in terms of infrastructure rests in imaginative thinking, which is also behind enframing technologies. This notion is consistent with Heidegger's (1993, 1997, p. 333) poetic ending of *The Question*, as mentioned previously: "But where danger is, grows[,] [t]he saving power also."

The saving power, reimagining the urban condition and transforming it, could benefit from an artistic approach. Such a measure should signify a new way of thinking that transcends enframing. As indicated above, there are already several approaches for dealing with infrastructure design, exhibiting thinking that is turning away from enframing. The notion of paramount importance for changing our mental life is to support measures and ideas that can save us from widescale ecological destruction and social injustice.

At this point in the section, we should have a general sense of what defines a saving technology for an urban environment. However, having a more focused account could benefit the accessibility of the terms' conception. In turn, establishing criteria could help guide their creation and implementation into cities. To that end, below is a non-exhaustive account that goes in that direction. It is important to note that it is indicative rather than absolute. One reason for this limitation is that defining saving technologies is a relatively nascent enterprise, meaning revisions or refinements could be in order. Making room for this consideration bolsters the view mentioned earlier that wicked problems require solutions beyond tame fields of study.

Second, flexibility in saving technologies could lend itself to situations wherein an urban technology aligns well with their spirit and aims. Still, it could fall outside of any strictly defined parameters. Rather than abandon support for a novel invention, reconsidering the theoretical framework could turn out to be a sound measure, one that could advance the thinking behind saving technologies. This notion does not suggest that we should stretch the definition of a saving technology to make room for devices that debatably qualify, but it does entail that some degree of flexibility could benefit efforts towards saving technology in the long run.

With this notion in mind, it is advantageous to understand how the pattern behind saving is applicable in several senses. In terms of saving thinking, it means contemplating short-term and long-term considerations that keep meaningful notions of respect in view while remaining outcome-dependent. The point worth underscoring here is that it describes a mindset geared towards developing a way of interacting that strives to keep standing reserve at a significant distance, engaging in the practice *extremely selectively* rather than seeing it as an absolute, default setting. This notion is qualified because being held in standing reserve suggests that respect is diminished or absent in such cases. While we can debate the degree of respect that humans, nonhumans, and artifacts deserve, another issue is examining what showing respect means for the person engaged in the actions that bring standing reserve into question. How or what does such thinking and actions say about he/she/them? Such thinking must also support worthwhile goals such as just urban sustainability and human flourishing.

Still, there could be cases wherein holding something in standing reserve would require dealing with conflicting groups of stakeholders to support the objectives above. Those views could collapse into a "lesser-of-evils" decision, which would demand thorough investigation to find the choice that would count as saving. Although this point leads to a separate inquiry for ethics—a different study in nature—the ontological question determined by the schema of wicked and saving remains. It concerns the character of thinking exclusively. It requires meaningful attention, serving as the starting point for designing a means of engagement with humans, nonhumans, and artifacts.

From the example above on saving thinking, we can extrapolate a pattern that pertains to how saving plays out in other contexts. For instance, the pattern behind saving thinking that informs how a technology affects humans, nonhumans, and artifacts should produce outcomes that reflect the degree of consideration and respect warranted in a particular situation while supporting worthwhile goals. While circumstances will change from one instance or city to the next, what remains constant is that outcomes must align with the description of saving. Experts of relevant stripes must be able to verify them. This point signifies that specialist across the academy, business, government, and affected residents must weigh in on the veracity of urban technologies deserving saving status.

Here is a list that puts the ideas above into a succinct perspective: (1) saving technologies should play a role in exposing people to positive mental states that improve their overall mental life. These states include but are not limited to love, joy, receiving care, and belonging. (2) They should not have detrimental impacts on other people or cause significant harm to the local, regional, or global nonhuman environment to the best of our knowledge. Such an idea also entails that they work well, e.g., align with (1), with other situated technologies. This point highlights that no saving technology is absolute and depends upon their outcomes in a given situation. A saving technology in Santa Barbara, California, could easily be a wicked one in Jefferson City, Missouri, or vice versa. We need to keep in mind that they must complement and improve existing structures while not starting, perpetuating, or exacerbating a social problem.

This point suggests that these technologies should align with environmentally and socially just sustainability, even though such accounts continue to progress. This point's motivation is to facilitate pathways to mental states that support the ideas in (1). (3) They should contribute to and represent the kind of thinking that sees people, nonhuman beings, and intrinsically valuable artifacts as significant. This notion means that any (nonhuman) entity should only be held in standing reserve if it is clear that doing so supports the previous points. Still, it slightly varies if it supports larger goals.

Yet, in a free relationship wherein we develop a way of thinking that allows for other forms of thought and associations, this category will require advanced scrutiny and debate. The reason here is that this criterion presents an opportunity for corruption and misuse. While adding this caveat at least creates the possibility that such an event might be less likely to emerge, not including such a provision makes more significant harm by excluding possible technologies that are the best of a worst-case scenario. Lastly, there is also the possibility that a completely new technology emerges that either replaces many technologies at once or shifts conversations such as the invention of social media that could offer benefits (that must be stacked against harms) that could count as another kind of saving technology.[4] Although the claims above sound rather ambitious, they face several challenges, which I explore in the section below.

4.3 Challenges to an Expanding Taxonomy

In Chap. 3, the challenges presented towards the end showed that specific issues of wicked technologies required attention. The same is true here. There are also problems with saving technologies that would benefit from further investigation. For instance, in the case above that establishes the ontological possibility for saving technologies, there is little concern for standing reserve, which leads to the questions: must saving technologies not hold nonhuman resources in standing reserve? Must we put distance between the mindset necessary to change and the thinking that got us here? The answer to both questions is no. Taming a wicked problem could require us to hold a natural resource in standing reserve to develop mitigatory efforts. The only caveat is that this kind of standing reserve does not produce significant environmental and social harm (or any injustice).

While this notion might appear to be inconsistent with saving power, Dreyfus (1997) argues that, for Heidegger, the point is not to romanticize revealing technology, only to gain command of it, along with enframing ones. Holding nature in standing reserve to work towards mitigatory measures does not reflect the mindset that would yield a narrow way of thinking that would be the only way to see the world. Conversely, thinking about creating a sustainable world, one where people can thrive and flourish, indicates that the primary motivation behind such endeavors is not driven by the desire to disregard nature or people. For instance, harvesting tons of biomass to provide energy might require viewing a plant species as a standing reserve. Yet, the thinking behind it is to develop solutions for mitigating harm and survival, not inherently producing "more" for its own sake, a point that is consistent with enframing.

In addition to these points, the model of thinking that we would associate with saving technologies should apply to all devices. Focusing on cities as whole entities and their smaller parts is an excellent way to advance our thinking on such matters. This notion makes sense when considering that their design and composition, including healthcare, water, energy, ICT infrastructures, food systems, public

---

[4] There is also the possibility that removing a technology, such as single-family zoning, could also support saving technologies or cities, which is an outlier scenario that could require further investigation.

spaces, housing, transportation, commercial districts, and industries, enormously impact the nonhuman world.

This reality shows that engineers, scientists, planners, researchers, and professionals who have the imaginative power to transform the world would, in turn, will also change themselves, exhibiting the reciprocal nature of such an enterprise. Due to this situation, urban technologies should receive advanced study, but we should also apply such measures to other entities that shape mental life. Bearing in mind that we are ultimately dealing with a way to improve our thinking, which affects the conditions for mental life, the above professions will lead by example. They can create the thought patterns that people need as examples to follow, leading to a complete overhauling of how we are thinking about technology and ourselves.

Looking back at the case of NOAs in southern Nevada, in terms of how professionals approached the issue, they engaged with a situation that could lend itself to future instances that involve workers' safety. Despite the reality that the bypass in Nevada facilitates the movement of fossil-fueled vehicles that pollute and that the land is held in standing reserve, which could be improved through better-quality technologies, this condition does not yield significant harm to people or the biosphere when considering that this effort could reduce emissions from gridlocked traffic from Las Vegas' bound travelers. This example illustrates the kind of watchfulness that one ought to associate with saving technologies as it matches the thinking described above, even though it could still be improved significantly.

Although this instance is an isolated case, it emblematizes the mindset necessary to save us from thinking in a manner that gave way to the myriad problems that emerged from the culmination of wicked technologies. Despite having limited effects, increasing the number of cases, when totaled, could have tremendous results. A piecemeal approach should not be underestimated when there are sufficient pieces to get the job done. This point suggests that we should not discount focusing attention on transforming urban technologies in hopes of changing, or saving, cities. Still, there is also the possibility that unforeseen technologies could improve life in ways. That is, at present, we cannot see how there is a need to do so in a given capacity because we are accustomed to our situatedness, but it or their invention could improve the urban condition such that they demand that we reexamine the concept of saving technologies or refine the taxonomy even more.

4.4 From Saving Technologies to Saving Cities (as Technologies)

While exploring these categories has previously helped us examine our technological lives, expanding this taxonomy could help us create a future worth living and the conditions for the mental lives that we want. This chapter identified how the taxonomy of technology that we inherited from Heidegger needs revisiting. After determining the limits of his insights, the nexus of infrastructure, urban sprawl, and

NOAs emblematize how socio-material arrangements hold new dimensions wherein humankind must consider the delayed harmful impacts that are inevitable in some instances. Yet, with creative thinking, developing "workarounds" can deliver solutions to mitigate harm. In turn, not pursuing technology with an unending enthusiasm for its own sake indicates a way of thinking that can improve the conditions for mental life, considering that it is moving away from the thinking that is consistently associated with enframing. Being mindful of the reality that wicked problems such as climate change now induce anxiety, which shapes the conditions for mental life, we understand the motivation to undertake such a task should hold steady.

Although climate change is one significant wicked problem that challenges cities, other wicked problems induce anxiety, along with different harmful situations that lack a helpful designation or description. For example, vast social and economic inequalities—expressions of racism, sexism, ableism, classism, or any other similar classification—plague the world's cities. There are also issues that we accept as customary conditions of urban dwelling: traffic, noise, pollution, crime, and corrupt public employees, to name a few. Either in isolation or combined, these matters can shape the conditions for mental life through inducing anxiety or fear. Yet, place-specific circumstances can also cause worry, including but not limited to fear of being homeless or losing one's home, displacement of the home, food insecurity, unstable transportation, unemployment, dying alone, or simply lacking essential resources.

The point here is that the arrangements of structures and the availability of services in the urban spheres create the contours of people's lived experiences, which are the primary ways to deal with the unique condition of having a mental life. Although the situations above play a significant role in creating the conditions that lead to anxiety and worse, each city will differ due to its unique historical, environmental, and political situatedness. Still, the pattern behind the conditions will produce similar manifestations that provide common ground for conversation and a starting point to examine the nature of such affairs to begin charting a way forward to help create mental states other than anxiety. They include but are not limited to love, joy, happiness, community pride, and urban comradery. These feelings can also make one aware of their mental life, but they do not come with harmful elements associated with anxiety, fear, or apprehension. If a city can make people feel anxious, it can also help them experience joy. Claiming otherwise shows a biased penchant for suffering, a demented view that could require medical attention.

Starting from the position that we can restructure or build cities that can give people feelings of love and or joy, this attitude can move us in the direction that we can drastically reconceive the foundations of thinking that influence the conditions for the quality of one's mental life. In turn, we learn how philosophical analysis could complement the inner workings of city halls so that we could move from "saving technologies" to "saving cities." The trick here is to figure out ways to upgrade urban environments, piecemeal in most cases, considering that we cannot simply start over, which is the book's direction.

This chapter focused on the needed mindset to deliver the required outcomes. The following chapter explores how other wicked problems such as those touched on above need it, also. The hope is that considering that cities and the climate will continue to change, we must always understand that the designation "saving city" requires attending to the conditions surrounding it. The term remains in a "towards" mode, moving toward the designation of saving city as a process rather than an achievable goal. Referring to a city that pursues this status with gusto entails that it cannot *really* be a saving city, but it can embark on the process of seeking such a classification, one saving technology at a time.

4.5 Conclusion

This chapter's essential takeaway is that shifting our thinking from one that focuses mainly on enframing and standing reserve to one that is advantageous to humankind's longevity is crucial towards developing saving technologies and saving cities. While such a move might sound straightforward, the reality is that such practices are deeply engrained in habits and ways of living. This notion means that transforming the world means starting with ourselves. The beneficial dimension worth noting here is that the people behind the schools of thought mentioned in the earlier sections (e.g., cradle-to-cradle) exhibit the kind of contributions required for an undertaking with no end. Yet, the reward serves us well, considering that taking such bold steps can move us toward securing better mental lives and our long-term survival.

The good news is that considering that we are dealing with a pattern of thinking with a wide scale of applicability, one that concerns mental life in general, one can expect that extrapolations could be forthcoming if such insights are maximized. In turn, a picture comes into view showing that despite the differing character of such affairs, the mitigatory mindset has enough utility to assist in those matters. The following chapter does just that. It illustrates that there are similar wicked problems and other troubles that demand an investigation. The efforts to deal with them effectively also rest on developing an outlook that moves towards saving.

One could make the case that the attention should focus on developing smart cities, resilient cities, or sustainable cities to save us. While those terms are useful, they are merely co-extensive designations. Similar to how "cohousing communities" can qualify as saving technologies, "sustainable city," "resilient city," "smart city," and "just city" could count as saving technologies, too. These terms could be helpful as thought technologies. However, we should have some reservations when applying such a label. They remain subject to how they function with thought technologies of various sizes and scopes, an idea that is consistent with the caveat that wicked problems are interrelated. This notion requires us to address two worries.

First, there are concerns about the terms' integrity and accuracy. Defining them illustrates their soundness and or the condition required for them to deserve such a categorization. Suppose the conception of a "sustainable city" is doomed because it is impossible to balance its pillars. In that case, one could argue that it should be discarded—unless one can account for such a condition in a way that dismisses such a concern. Second, as with all saving technologies, they should only receive the designation if their saving status is achieved while working with other technologies.

Still, when employed fittingly with other socio-material entities, such conceptions can contribute needed efforts to create saving cities, and the following chapter fleshes out this process.

References

Dreyfus H (1997) Heidegger on gaining a free relation to technology. In: Shrader-Frechette K, Westra L (eds) Technology and values. Rowman & Littlefield Publishers, Lanham, MD, pp 41–54

Farr D (2011) Sustainable urbanism: urban design with nature. Wiley, Hoboken, NJ

Friedman B, Hendry D (2019) Value sensitive design: shaping technology with moral imagination. MIT Press, Cambridge, MA

Heidegger M (1993, 1997) The question concerning technology. In: Krell DF (eds) Martin Heidegger: basic writings. HarperCollins Publishers, New York, pp 311–341

Leopold A (1949, 1968) A sand county almanac: sketches here and there. Oxford University Press, New York

McDonough W, Braungart M (2010) Cradle to cradle: remaking the way we make things. North Point Press, New York

Meerow S, Newell J, Stults M (2016) Defining urban resilience: a review. Landsc Urban Plan 147:38–49

Tracada E, Caperna A (2013) A new paradigm for deep sustainability: biourbanism. In: The proceedings of international conference & exhibition' application of efficient & renewable energy technologies in low cost buildings and construction. 16th–18th September, Ankara, pp 367–381

Chapter 5
Saving Cities

Abstract Having established an account of saving technologies in the previous chapter, this one moves towards saving cities. This move's motivation is to show that if the former can help increase positive mental states (love, joy, etc.) that create better mental lives, we can work towards meaningfully giving the latter the same categorization. On the one hand, this position holds that having many saving technologies creates numerous opportunities for them to work in unison to achieve the outcome above. On the other hand, it shows that cities, when thought about as technologies, can qualify as saving technologies, also. In turn, it makes sense to call such places "saving cities", especially when considering that they could protect people from some immediate harm and humankind from an anthropogenic demise. To make this case, this chapter explores a somewhat novel housing technology called "cohousing." The purpose behind this move is to exhibit how to apply the structure of "saving" to a technology that can facilitate immediate efforts that can help shape mental life, which could also contribute to long-term strategies to do the same on a municipal scale.

Keywords Saving technology · Saving cities · Cohousing

5.1 Introduction

After providing the grounds for what counts as a saving technology in Chap. 4, establishing a view showing how these devices play roles in creating saving cities is now the aim. This undertaking is a two-fold exercise. First, it indicates increasing the number of saving technologies in the urban environment helps create positive mental states that enhance mental life. Such endeavors can replace existing ones with models that mitigate harmful mental states such as anxiety, fear, and stress.[1] Second, if we can accept (or entertain) the view that either gaining a substantial amount of saving technologies gives us a way to call a city a "saving city" or such devices compose much of the city, which requires us to see the city as a larger technology,

---

[1] This point is not meant to say that all such mental states are inherently bad because cases can be made showing that there are circumstances wherein they would be positive such as survival. To be clear, I am not talking about those cases.

then calling ones that align with the description established in the previous chapter "saving cities" makes sense.

To unpack these claims, this chapter begins by examining a part of the city required for urban dwelling: the home. It specifically investigates some of the harmful mental states commonly associated with homes as wicked technologies. This process allows us to see a few of the problematic elements entailed in how we think about the places we inhabit. From the outset, this notion is troubling because the home's mere conception should entail safety, security, and, hopefully, happiness. Yet, due to larger eco-socio-political forces that come with wicked problems, some contemporary homes and housing arrangements stop short of being hostile in some instances and can practically serve as torture devices in the worst cases. Still, they essentially qualify as wicked in most situations. This notion is especially significant if we consider a city's housing collectively, at least in places where the topic continuously causes controversy.

One technological (partial) remedy is cohousing, which, inherent in its conception, holds the potential to mitigate harm by providing meaningful access to positive mental states. Through exploring this kind of saving technology, we gain a model that emblematizes the sort of thinking that can help create desired places. This process then shows that we can create saving cities by employing this practice over a considerable duration. The chapter concludes by looking to the future for areas of study that could help advance these ideas.

5.2 Wicked Housing

There is no doubt that cities play a significant role in climate change, consuming 78% of the world's energy and produce more than 60% of the globe's greenhouse gases (United Nations n.d.). In the United States, residential buildings account for about 20% of their emissions (Goldstein et al. 2020). Additionally, cities can be exhausting, loud, and chaotic. Urban dwellers need a place to unwind, refresh, and protect themselves from the world external to them. The home is the place to do these activities or nothing at all. Although one could argue that homes should be relaxing for these reasons, the eco-socio-political conditions surrounding them can diminish their capacities and connects them to climate change.

Despite this global reality, local geography will largely determine the features required for curtailment. This point suggests that even though we are dealing with tame topics such as emissions that remain measurable, governing forces, history, and socio-cultural customs hold steady in such affairs. Zoning, for instance, largely determines what kind of housing developers can build (Manville et al. 2020). Urban growth boundaries can promote density in some cases (Gennaio et al. 2009). Minimum requirements for parking spaces play a significant role, taking up limited and valuable real estate (Belmore 2019). As mentioned previously, some cities' design helps create the conditions for "forced car ownership", wherein life becomes exceedingly challenging without a personal vehicle (Currie & Senbergs 2007). There is also the fear of being displaced from one's home due to actualities such as gentrification. The mere struggle to make ends meet, forcing people to deal with balancing housing finances with other monetary demands, can remain a constant source of worry.

Regarding vulnerable groups, such conditions have contributed to the public-health tragedy of social isolationism, a condition that could make seniors suffer lonely deaths (Holt-Lunstad et al. 2015).[2] Imagine, for example, that you have raised your children. They moved away to find work and cannot visit often. Your spouse dies. Having spent a lifetime working, you do not have many friends, and the friends that you do have are suffering from similar conditions because they are similar in age. They are dying, one by one. The nights of getting together for dinner and drinks are memories. Even though you can still move about your house and make trips to secure provisions, most of the people you encounter are due to monetary exchanges (recall Simmel), and the pleasantries that they afford are quick and empty. You wish there were community centers or events for seniors, but you live in a place where the term "community" is disliked because it takes away from the reward of pulling yourself up by your bootstraps—or something to that effect. Due to these or similar conditions, the loneliness that you experience is undeniable. It is always there. You now long for the daily humdrum of your old life, even the annoying people—at least there were people present who could get on your nerves. There is also the fear of slip-and-fall injuries, choking to death, and due to greed and evil, you worry about "helpful" guardians who want to stick you in facilities where Nurse Rachet would be Employee of the Month.

While I wish that the notions above were hyperbole, previous cases show otherwise (Pogach & Wood 2019). It is easy to dismiss this issue if it does not apply to *you* (yet), and one can quickly reason such worries away because *it will not happen to you.* While there is no way to disprove such a view, it seems challenging to fathom that the people in precarious positions as described above planned for them. While arguing that each senior suffering from loneliness is responsible for her or his happiness is a view that one *could* take, embracing such an outlook makes you the kind of person who lacks a shred of empathy, arguably. Why should such a person's opinion be meaningfully considered in a society that wants to show respect and love for each other?

It should not, and such views should be outright rejected due to their ridiculous nature. Yet, the notion that someone would make that argument hints at more significant problems, such as how morally bankrupt some societies have become. As stated in the previous chapter, wicked problems are often connected to larger wicked problems, and issues of the home are exemplars. Even though larger issues require attention, suffering people still need help. Arguing that we should not devote *some* thinking, support, or actively work to triage people's harmful mental states for

---

[2] For more information on social isolation as a public-health issue, see Cacioppo and Cacioppo (2014) Social relationships and health: The toxic effects of perceived social isolation." *Social and personality psychology compass* 8(2) pp. 58–72. Also, Holt-Lunstad et al. (2015) Loneliness and social isolation as risk factors for mortality: a meta-analytic review." *Perspectives on psychological science* 10(2): pp. 227–237. Also, see Cotterell et al. (2018) Preventing social isolation in older people. Maturitas, 113, pp. 80–84.

the sake of fighting against neoliberalism or bringing down the capitalist system is "luxurious" and a convenient excuse not to do anything.

Moreover, it could perpetuate *actual* harm, considering that a successful hindering of such a view could eliminate efforts to stop seniors from facing lonely deaths. Granted, my view says nothing about ending the struggle for global, national, state, or municipal justice—but it does not actively champion or stymie such measures. There is no good reason why *most* researchers cannot continue their studies regarding the issue above while others work to combat the problems related to harmful mental states. This notion is axiomatically underscored by the reality that the above issue concerns—housing—a necessity for existence. Shelter is the first thing that survival experts say that one should secure when stuck in the wild without anything else (Macwelch 2019).

The point here is that we can isolate more significant issues as they are entangled in more encompassing affairs to help facilitate people's exposure to positive mental states that create rich mental lives, which can mitigate harm and replace existing wicked technologies in metropolitan places. One problem is that the nature of wicked problems holds that there are no concrete solutions, suggesting that a one-size-fits-all approach is unlikely to find success. In turn, developing specialized measures to address housing situations on a case-by-case basis appears to have the kind of orientation required to deal with troubling scenarios such as those mentioned above.

Still, efforts to alleviate these concerns should aim to remedy as many undesired elements as possible. For instance, when proposing a corrective avenue for social isolationism, it should also bring considerations for just sustainability into view. Although this example accounts for a particular matter, it embodies the kind of thinking consistent with forming a free relationship with technology, one that has saving qualities. The following section investigates how a relatively novel technology known as "cohousing" inherently speaks to the worries above to illustrate these points. It shows that, while imperfect, in the appropriate setting, they can help improve the wicked problem of housing, delivering a kind of technology that either qualifies as saving or leans in that direction, which could count as progress toward saving cities, albeit in a piecemeal fashion.

5.3 Cohousing as Saving Housing

By definition, "cohousing" is shorthand for "co-operative housing" (Wang et al. 2020, p. 248). It means that housing is mutual, but this does not exclusively entail that people live together (McCamant & Durrett 1994). Instead, cohousing arrangements can take several forms. The central idea is that these are *intentional* communities (Choi 2008; Boyer & Leland 2018). Residents plan to live this way for collective aims that involve some degree or kind of social elements, which can vary from one community to the next (Wang et al. 2020). Some groups might share views on the environment or politics, but those aspects are typically coextensive qualities that are not required or essential (Boyer & Leland 2018). In essence, people or individuals in cohousing have their private residences or living quarters, but shared space is utilized (McCamant & Durrett 1994).

Examples could include dining areas, kitchens, courtyards, gardening space, and recreation areas (McCamant & Durrett 1994). They frequently eat meals together, carpool, exercise, share resources, and engage in activities wherein there is mutual interest (McCamant & Durrett 1994; Wang et al. 2020). In many cohousing communities, the composition is multigenerational (McCamant & Durrett 1994). This point suggests that childcare could be built in, seniors have company, and their wisdom could benefit the group. The inherent structure has room for social connectivity, eliminating the wicked condition associated with social isolationism. To this end, we see that cohousing communities have an enhanced social dynamic unmatched by traditional multi-family housing operations such as apartment complexes and condominiums.

While cohousing communities provide numerous benefits for the people who participate in intentional living lifestyles, they also could arguably make a measurable impact on reducing carbon emissions (Wang et al. 2020). This notion suggests that cohousing offers significant advantages for urban areas that face challenges such as those described in the previous section. Even though cohousing is somewhat unknown by many mainstream populations, studies show substantial impacts that warrant further investigations (Wang et al. 2020).

For example, they reduce greenhouse emissions (*ibid.*). Reduce purchase through sharing of household goods (*ibid.*). Food waste is reduced (*ibid.*). Some cohousing communities carpool to maximize fuel usage (Beck 2020). Some residences fare better in terms of reduced use of electricity, and the trend here is that cohousing groups use less power than traditional housing (Brown 2004). When considering these positive benefits in tandem with the social connectivity elements mentioned above, they stack up well against the criteria for a saving technology.

Recall these points from the previous chapter. Saving technologies need to expose people to positive mental states that help them have good mental lives. Such mental states include feelings such as love, joy, and belonging. They should not cause significant harm to the nonhuman environment, meaning that these technologies should align with environmentally and socially just sustainability, even though such accounts continue to require progress in certain respects. For instance, not all cohousing communities align with just sustainability, but it is reasonable to hold that any nascent enterprise should be allowed to self-correct while establishing itself (Epting 2018).

This point aside, although cohousing communities have an established history, they remain on the outside of the mainstream in most instances, having faced much discrimination from financial institutions and government entities (McCamant & Durrett 1994). Due to their inherent nature that offers an alternative to single-family housing, often being viewed as hippie communes, these initiatives must contend with and overcome prejudices (Lockyer 2007; Boyer & Leland 2018). Despite such hurdles, they remain steadfast in their endeavors, finding unique workarounds to exist as communities (McCamant & Durrett 1994). In turn, cohousing communities exist globally, and they have established online networks to assist each other in relevant matters (Cohousing 2021).

Briefly put, their design principles and real-world impacts align well with saving technologies, showing how they can replace wicked ones to help deliver better urban futures. The point to keep in mind is that cohousing communities are not the solution to all housing troubles, but they can be significantly helpful under the right conditions. The essential idea that must remain at the forefront of our thinking is that they embody principles that move away from enframing and wicked problems. It shows that developing a free relationship with technology means that when improved options exist, choosing them suggests that we are not limited to the kind of thinking that only looks at the world and people as standing reserve. While modest, this sort of thinking over time could have a meaningful effect if similar enterprises complement it. Increasing the number of saving technologies in cities could provide more ways for urban dwellers to experience positive mental states. This idea is the motivation behind saving technology in the urban sphere, and the following section fleshes out this notion.

5.4 Saving Cities in Focus

As mentioned, cities consume most of the world's resources and hold steady as sites of vast social inequality—platforms for wicked problems that complicate mental states as outlined previously. Regarding the environmental impacts, those tame aspects are measurable. Research areas such as industrial and urban ecology make that point evident. However, wicked elements include concerns that challenge the ontological structure of cities and the nonhuman world. For instance, in his seminal text, *The Imperative of Responsibility: In Search of an Ethics for the Technological Age*, Jonas (1984) situates the city as an enclave, stacking it against the nonhuman world to illustrate the changed nature of technology and its longstanding impacts. Jones (1984, p. 10) argues:

> For the boundary between 'city' and 'nature' has been obliterated: the city of men, once an enclave in the nonhuman world, spreads over the whole terrestrial nature and usurps its place. The difference between the artificial and the natural has vanished, the natural is swallowed up in the sphere of the artificial, and at the same time the total (the works of man that have become 'the world' and as such envelop their markers) generates a 'nature' of its own, that is, a necessity with which human freedom has to cope in an entirely new sense. Issues never legislated come into the purview of the laws which the total city must give itself so that there will be a world for the generations of [humans] to come.

One way to employ this passage's wisdom is to pair it with tame elements examined earlier, showing that cities consume substantial resources from the nonhuman world. They are sophisticated mediums in two senses. On the one hand, the process of materials flowing in and out of cities puts significant demands on the nonhuman world (Hodson et al. 2012). One could reasonably argue that these exchanges remain out of the immediate view of many urban dwellers. Still, they depend on it for existence, taking resources away from the nonhuman world while diminishing it. When thought about as technologies, this view shows that cities are wicked to the extreme, even though this point concerns *only* their relationship with the nonhuman world. The worry here is that this process obfuscates cities' consumption, making it challenging to see the connection, compromising humankind's ability to contemplate myriad issues that intersect with this notion.

On the other hand, looking at how many cities continue to expand horizontally shows that appropriating additional land inconspicuously increases urban landmass while diminishing space for flora and fauna. Such measures will benefit numerous people, meaning that there is little reason to challenge it, except for people who have respect and sympathies for the nonhuman world. This latter point brings the need to deal with wicked urban issues into our thinking. The problem remains definitively urban, considering that municipalities are the ones who are working to secure additional housing, and the nonhuman world remains relegated as standing reserve.

The point here is not to vilify urban sprawl, only to illustrate that there is little concern for developing a solution to urban housing that embodies problem-solving measures that bring the advanced thinking behind saving technologies into the picture. By engaging in such practices instead of choosing the path of least resistance, cities can solve a problem that exists now, such as the lack of housing. Yet, they can also prevent additional ones in the near and distant futures, such as the nonhuman world's diminished capacity, which brings unknown and unforeseen elements into view.

One helpful way to think through such affairs is to think about cities as technologies, despite such a task's complicated nature. Elsewhere, I argued that when examining urban thought literature throughout several decades, a "received view" emerges, showing that it is common to think about cities as technologies (Epting 2021). According to the previous chapter's description, some such positions align well with Heidegger's account of an enframing technology, and they qualify as wicked ones. One excellent example is Félix Guattari's (1987) insightful use of Lewis Mumford's term "mega-machine," which involves merging technology, political influence, and science (Genosko 2015). Here is how Guattari (1987, 105) puts it:

> Cities have become giant machines, "megamachines," to borrow an expression from Lewis Mumford, producers of individual and collective processes of subjectivation by means of collective apparatuses (educational, health, social control culture) and mass media. The material infrastructure, communications and services of cities cannot be separated from functions that may be described as existential. The urban drama being played out on the horizon at the end of the millennium is only one aspect of a crises much more fundamental that threatens the future of humankind around the planet. Without a radical reorientation of methods, and particularly goals of production, the global biosphere will be thrown out of balance and develop toward a state of total incompatibility with human life and all forms of animal and vegetable life in general.

From this passage, we see that the idea of the wicked city aligns well with Guattari's thinking on the topic. Standing reserve goes without saying in the context above. It is embedded in the conception of the city as a megamachine, turning humans into resources. In turn, people exist to serve the city—rather than the city serving as a way to perpetuate humankind's permanence and make our wildest dreams come true.

Mental states become inconsequential, meaning that we must invert the process to secure positive ones. What sense does it make to serve a machine—unless we redefine the term to eliminate the view above. Here, we find the vital impetus for saving cities. After identifying, examining, and making a case against wicked technologies in the city, the next step is developing or locating saving alternatives. Repeating this process time and again will create more opportunities for people to experience positive mental states while decreasing negative ones, suggesting that they could create desired mental lives under favorable circumstances. The point worth keeping in mind is that progress, not perfection, is the goal for such an eternal undertaking.

This notion is consistent with the character of wicked problems. We cannot solve them, but we can take meaningful steps to tame them to a manageable degree to live with them. We are always involved in the process of saving because they perpetually change. Moving away from wicked to saving, replacing the conception of the city as a megamachine, entails pursuing a free relationship with the city.

The good news here is that strategically reconceptualizing cities as technologies that include this consideration provide the means to save us from the dire situations that accompany the urban condition. In turn, thinking about metropolitan environs geared in this direction can rescue humanity from imperilment. By clarifying the significant role that cities play in shaping our mental lives, urban dwellers should help shape them.

However, such a task on practical levels will require efforts from numerous community members and specialists, signaling that there are several areas of research to make effective change. This process will be long-lasting, and it will require thorough examination and regular maintenance. Although pinpointing all topics relevant to this sort of undertaking remains an inherently daunting enterprise, the section below brings some pertinent subjects into view. When thinking about the shared thread that bounds them together, the significant notion is how they affect mental life.

In turn, identifying the ones that maintain a prominent position in cities should hold steady as a guide for determining which wicked technologies should receive attention quickly. Keeping in mind that cities differ, elements that align with the descriptions laid out in previous chapters will also. Yet, cities often bear a family resemblance, meaning that they can and do learn from each other—even though concerns for how to design, create, implement, modify, or upgrade urban technologies to achieve saving status is probably not a topic of discussion at City Hall. Still, even though urban professionals are probably not aware of the subject does not mean that the realities that correlate to its effects do not require attention. We are dealing with mental life and how the city helps shape it. It might not seem evident at first that this affair intersects with the city. Still, with over half of the world's populations living in urban environments, it should not come as a surprise that we should study how they affect fundamental topics that concern the quality of our mental lives.

5.5 Towards Future Areas of Research

While this chapter used cohousing as an example of a possible saving technology, it is not the only way to replace housing situations that qualify as wicked. Other means that achieve similar results unique to places are needed to alleviate harms associated with negative mental states. As they become known and their success is reproduced, studying them could be advantageous for cities with the same kind of housing trouble. Considering that urban residents require homes to dwell in, making progress should be a constant goal. Forming free relationships with housing technologies and new and emerging smart devices can provide people with the means to lessen the demand for standing reserves in some instances.

Transportation is another area that dramatically needs saving: traffic jams, excessively long commute times, pedestrian and bicyclist fatalities, pollution, and global climate change are a few of the significant issues associated with navigating the cityscape (Epting 2021). If one urban strand calls for attention to move away from wicked technologies globally, this one qualifies. Although industry leaders proselytize a future for automated vehicles or bust, there is no guarantee that these devices will count as saving technologies, despite having significant promise.[3] Other options such as light rail, bus rapid transit, ariel streetcars, and or multimodal transport systems could produce better outcomes for mental life. Yet, these options could also be ineffective compared to the realities that car-free or reduced-car urban mobility could produce in some cases. This option could save numerous cities.

Again, the examples above are merely illustrative. Long-term actualities could exhibit that other devices fared better regarding the conditions that they help create. Still, they indicate the kind of thinking that could underpin cities' ontological structures geared towards humankind's longstanding survival. It is then no exaggeration to say that to exist as beings who favor a good mental life, urban technologies should be partial towards this ultimate endeavor. Recalling a point made earlier in Chaps. 3 and 4, if methodological thinking behind saving power qualifies as a technology, then moving towards saving technology starts there. Securing this mindset is also subject to careful examination for the same ends. This condition is not a weakness of the saving power. It signals the strength and utility that accompanies the need to adapt to changing scenarios.

To grasp its potential, the following chapter investigates the technologies mentioned above and others that embody the saving spirit as laid out thus far. A pattern comes into view by having a few more examples that provide an enhanced sense of how the identified requirements for saving technologies could benefit residents throughout the city. Although determining how to secure such conditions is an ambitious undertaking, gaining a more precise picture shows how to achieve it, despite its daunting nature. Still, even though it is doubtful that we can stop wicked problems from affecting cities, including saving ones can move toward "balancing"

---

[3] To see industry leaders who hold such views, see Allegro, J (2020) How Google's Self-Driving Car Will Change Everything. *Investopedia* December 20. https://www.investopedia.com/articles/investing/052014/how-googles-selfdriving-car-will-change-everything.asp.

harmful effects, even though stability does not guarantee any substantial benefits. The hope is that achieving some degree of equilibrium puts humankind in an advantageous position, and the innovative spirit continues.

5.6 Conclusion

This chapter explored the idea that we can create saving cities by pursuing saving technologies, examining how cohousing communities can serve as exemplars under fitting conditions. By providing a concrete illustration of how such a technology could help mitigate harmful conditions surrounding traditional technologies that, due to their eco-socio-political situatedness, are wicked. The purpose behind this move was to show that other urban worlds are possible, which we can build with a saving mindset and some elbow grease. To show that this kind of thinking and envisioning holds the possibility for sustained longevity, future areas for further study were presented. These topics listed were not exhaustive, but they signaled the sort of similar directions that leaders and innovators could explore to deliver, secure, and maintain future cities.

These pursuits are worthwhile because they assure our continued survival on a planet that withstands our technological successes. They exhibit that we can advance the mind-frames that define humankind, a perpetual task wherein progress is the goal. It is the guiding spirit that must hold steady as we continue technological undertakings, but it also must help define advancement. It is not the aim for which other elements hinge upon for their deserved place in humankind's affairs—but replaces those determining such measures surreptitiously. In turn, the outcomes are not the prime focus in the conventional sense. Instead, thinking as the methodological practice that seeks transformation as an inherent process must hold our attention. With this notion in mind, replacing perfection with progress is the pattern that aligns more smoothly in a world in flux.

References

Allegro J (2020) How google's self-driving car will change everything. Investopedia December 20. https://www.investopedia.com/articles/investing/052014/how-googles-selfdriving-car-will-change-everything.asp

Beck A (2020) What is cohousing? Developing a conceptual framework from the studies of Danish intergenerational cohousing. Hous Theory Soc 37(1):40–64

Belmore B (2019) Rethinking parking minimums. Inst Trans Eng ITE J 89(2):4–4

Boyer R, Leland S (2018) Cohousing for whom? Survey evidence to support the diffusion of socially and spatially integrated housing in the United States. Hous Policy Debate 28(5):653–667

Brown J (2004) Comparative analysis of energy consumption trends in cohousing and alternate housing arrangements. Doctoral dissertation, Massachusetts Institute of Technology

Cacioppo J, Cacioppo S (2014) Social relationships and health: the toxic effects of perceived social isolation. Soc Pers Psychol Compass 8(2):58–72

Choi J (2008) Characteristics of community life in foreign intentional communities focus on the differences between ecovillage and cohousing. Int J Hum Ecol 9(2):93–105

Cohousing (2021) Cohousing. https://www.cohousing.org

Cotterell N, Buffel T, Phillipson C (2018) Preventing social isolation in older people. Maturitas 113:80–84

Currie G, Senbergs Z (2007) Exploring forced car ownership in metropolitan Melbourne. Australasian Trans Research Forum 2007

Epting S (2018) Cohousing, environmental justice, and urban sustainability. Environ Ethics 40(2):135–151

Epting S (2021) On municipalities as technologies. Philos Technol. https://doi.org/10.1007/s13347-020-00438-z

Epting S (2021/forthcoming) The morality of urban mobility: technology and philosophy of the city. Rowman and Littlefield International, London

Gennaio M, Hersperger A, Bürgi M (2009) Containing urban sprawl—evaluating effectiveness of urban growth boundaries set by the Swiss Land Use Plan. Land Use Policy 26(2):224–232

Genosko G (2015) Megamachines: from Mumford to Guattari. Explor Media Ecol 14(1–2):7–20

Goldstein B, Gounaridis D, Newell J (2020) The carbon footprint of household energy use in the United States. Proc Natl Acad Sci 117(32):19122–19130

Guattari, F. (2015). Machinic eros: writings on japan. In: G. Genosko and J. Hetrick (eds) Univocal Publishing, Minneapolis

Hodson M, Marvin S, Robinson B, Swilling M (2012) Reshaping urban infrastructure: material flow analysis and transitions analysis in an urban context. J Ind Ecol 16(6):789–800

Holt-Lunstad J, Smith T, Baker M, Harris T, Stephenson D (2015) Loneliness and social isolation as risk factors for mortality: a meta-analytic review. Perspect Psychol Sci 10(2):227-237

Jonas H (1984) The imperative of responsibility. In search of an ethics for the technological age. Chicago University Press, Chicago

Lockyer J (2007) Sustainability and utopianism: an ethnography of cultural critique in contemporary intentional communities. The University of Georgia, Athens, Georgia

Macwelch T (2019) Survival shelters: 15 best designs and how to build them. OutdoorLife. https://www.outdoorlife.com/survival-shelters-15-best-designs-wilderness-shelters/

Manville M, Monkkonen P, Lens M (2020) It's time to end single-family zoning. J Am Plann Assoc 86(1):106–112

McCamant K, Durrett C (1994) Creating cohousing: building sustainable communities, New Society Publishers, Gabriola Island, British Columbia

Pogach D, Wood E (2019) When the guardian is an abuser. https://ncler.acl.gov/getattachment/Legal-Training/When-the-Guardian-is-an-Abuser-Ch-Summary.pdf.aspx?lang=en-US

United Nations (No Date) Cities and pollution. https://www.un.org/en/climatechange/climate-solutions/cities-pollution

Wang J, Hadjri K, Bennett S, Morris D (2020) The role of cohousing in social communication and sustainable living environments. WIT Trans Built Environ 193:247–258

Chapter 6
Saving Cities: The Road Ahead

Abstract While the preceding chapters laid out several of the theoretical and practical elements that come with saving technologies, the view showing how to create saving cities would benefit from additional explorations. This chapter does just that. It identifies some wicked technologies, and then it examines some nascent alternatives that could help municipalities move towards saving-city status. In turn, a more comprehensive picture emerges, revealing more of what a saving city would look like. Although these technologies will require additional research and development in city halls, voting booths, laboratories, and boardrooms, the conceptual depictions below provide motivation and guidance for such measures.

Keywords Saving technology · Saving cities · Future research

6.1 Introduction

Implementing a cohousing community into an urban environment shows how a saving technology can help us replace wicked ones. It serves as a model that demonstrates the elements such as environmental and social impacts that require improvement. Integrating one is not a highly efficient contribution towards mitigating climate change's effects. Yet, it is a significant component in the sense that, as a building block, its integrity has fewer internal tensions than wicked housing models do. For each cohousing community that connects people and reduces resource use, each instance wherein a wicked technology is replaced counts as a step in the right direction, moving towards a free relationship with technology with positive effects. It might seem counterintuitive to replace what exists, bearing in mind that wicked technologies already fulfill roles in our lives. On the surface, such a position makes sense. However, the targeted structures that must change are more than homes, vehicles, and codes.

Their principles rest on unstable ground. They used to guide our affairs soundly but knowing that their situatedness shows that they can no longer stand as we once understood them. Holding onto them due to their familiarity remains shakier than their foundation. Views underestimating or failing to address technologies' far-reaching impacts on people and the planet make such enterprises unreliable. Continuing to

S. Epting, *Saving Cities*, SpringerBriefs in Philosophy,
https://doi.org/10.1007/978-3-030-85833-9_6

depend on them when we can develop thought technologies betrays the innovative spirit that accompanies humankind's most novel achievements.

Letting them go so that we can replace them soundly is as brave as it is necessary. If we take the tenets of saving seriously, we gain a perspective showing that such principles are inherently enduring. This long-lasting quality means that any perceived waste is null in broad sustainability equations. When we gauge the longevity of the re-principled enterprises, temporary losses can support long-term gains. We will not only replace physical technologies such as inefficient housing that wastes resources and abstract technologies such as zoning ordinances that discriminate against neighbors, but we will replace the technologies of thought that built them.

This notion is where this chapter begins. It examines a few thought technologies, illustrating how they can replace ones that qualify as wicked in select instances. Next, it explores a few novel ideas that signal success in the real world or embody a beneficial spirit to help secure the conditions for saving cities. The chapter concludes by looking at the additional steps required to save cities ethically. Although this aspect raises further questions, due diligence can help us transform cities without seeing them as ends that justify wrongs. As Dr. Martin Luther King Jr. (1963, para 35) reminds us: "the means we use must be as pure as the ends we seek."

6.2 Saving Thought Models as Technologies

As mentioned previously, smart cities, sustainable cities, and resilient cities can serve as thought technologies. While researchers have paid much attention to defining and debating them, future inquiries that use saving technologies could examine their (competing) roles within larger systems to determine if they will assist in the collective efforts that can save us. This point also includes seeing how, or if, they can work in unison, in tandem, or any complementary fashion. The significance here is that each term could have unique functions that their interplay could reveal, further bolstering mitigatory efforts.

Considering these terms as thought technologies also means that they could house any number of smaller thought technologies—as they could also help compose larger ones. For example, if "smart city" is a larger technology, then information-communication technology could also be a smaller thought technology that makes employing any number of smaller technologies easier. However, the smaller ones would not hold a specialized use in many instances. This view halts an infinite regress of thought technologies, and it shows that there limits that would otherwise push the notion into absurdity.

The point here is to move on to examining how thought technologies such as information-communication technologies can work with tangible entities such as satellites to help create saving cities—remembering that by thinking *with* saving thought technologies—we are *still* engaging in a free relationship with technology. This point exhibits that rather than at the computer, laboratory, or drawing board, approaching saving technology starts in our minds as methodological thought geared towards creating the conditions for saving technologies and cities.

In turn, the technologies examined below are simple suggestions, which might or might not benefit saving cities. This point does not suggest that looking at them is a waste of time. It means that we should not endorse a saving technology until we consider it in a specific location. Recalling from when we saw the criteria for a wicked problem, we cannot simply install a highway to see how it fits in with other urban elements. However, with the best scientific, engineering, and policy minds working in concert, reducing some troubles appears probable. This point suggests that one avenue for future research should focus on finding commonalities between saving technologies to provide shared grounds for conversation, meaning that researchers could learn from each other. The examples below collectively illustrate the importance of bringing the notions above into our perspective of saving technologies as they relate to saving cities.

6.2.1 Aerial Cable Cars

Several cities view increasing the number of traffic lanes as the solution to having too many vehicles on the road. As a result, many urban environments have sacrificed prime real estate—only to create more traffic jams, increase pollution, harm human health, destroy historic neighborhoods, and commit unjust acts against marginalized people (Bullard & Johnson 1997; Epting 2021). Experts continue to demonstrate how such measures do not work and can make the problem worse in many instances. Some planners hold that alternative solutions such as diversifying mobility modes are a far more effective approach (Spickermann et al. 2014). While many cities are similar, each has unique topographical, historical, cultural, political, and economic issues to consider when addressing transportation affairs. In turn, one could argue that there is no one-size-fits-all approach to the mobility challenges that cities face. Still, the significant notion worth pointing out is that, despite the differences above, the common factor is automobile-centric places in numerous instances.

When faced with the reality of dealing with such matters, cities such as La Paz, Bolivia, Caracas, Venezuela, and Medellin, Columbia embraced a novel measure that could qualify as a saving technology for them: cable propelled transit systems, otherwise known as aerial cable cars and urban gondolas (Tischler & Mailer 2019). These cities have incredibly steep terrain, and researchers show that these technologies successfully overcome this challenge to effectively transport people, becoming mainstays in transportation systems with positive effects for users (Tischler & Mailer 2019; Garsous et al. 2019).

In the context of saving technologies that contribute to saving cities, these ariel cable cars, while reducing fossil fuel consumption, decrease travel time by 22% on average in some instances (Garsous et al. 2019; Inclusive Infrastructure 2021). Research on their impacts in Medellin shows that they significantly help marginalized communities (Matsuyuki et al. 2020). In La Paz, the *Mi Teleférico* (translation:

my cable car) exemplifies how a saving technology can improve people's mental lives. Before its implementation as the transportation system's mainstay, residents depended on private vehicles and buses for mobility through challenging terrain, leaving them with mobility needs unfilled or grossly inadequate, and financially burdensome (Inclusive Infrastructure 2021). Today, *Mi Teleférico* is an incredible success in Latin America, accumulating a surplus of almost 6 million USD (*ibid.*). Yet, it is surprisingly affordable with a standard fare of 3 *bolivianos* [0.43 USD] (Moss 2018). It is known for its focus on the needs of disabled persons and people who have impaired mobility, seniors, and students who only pay half (Inclusive Infrastructure 2021).

These elements show how these novel transportation options can qualify as saving technologies in these cities. However, they have drawbacks, meaning that urban gondolas would not be a feasible option for many places. This notion underscores the caveat that mitigatory efforts must remain location-specific. The goal here is to show how this kind of thinking can mitigate harms from outdated transportation systems, and it can also reshape cities and urban mobility experience. In turn, the anxiety and stress typically associated with traveling through these topographies are reduced, contributing to beneficial efforts for residents' mental lives.

The takeaway here is that transportation planners, engineers, and municipal officials can study technologies that work in particular instances to see if they will function properly or be modified to fit in with different scenarios to combat similar problems. Yet, the case above is a physical technology. In other situations, abstract technologies can also help saving cities. Despite this ontological difference, both align well with the thought model of wicked and saving. Consider, for example, that extrapolating from democratic practices can reveal the positive dimensions that, with refinement, can combat wicked problems and technologies. In the section below, I examine how one such novel technology does just that.

6.2.2 Participatory Budgeting

While most technological advancements that favor saving technologies as they contribute to saving cities are physical ones, some are more abstract. They are things such as building codes, regulations, and practices. One that fits this category at the municipal level is participatory budgeting. This democratic technology had its humble beginnings in Porto Alegra, Brazil, in 1989 (Smith 2009). It is a process wherein community members work in groups to determine feasible neighborhood projects that require funding, vote on them, and then work with municipalities to bring them to reality (Wampler 2012; Menser 2012; Epting 2020). In many instances, municipalities cover expenses through using discretionary funds (Menser 2012). Some examples include community gardens, transportation safety measures, library initiatives, and feeding seniors with innovative programs (Participatory Budgeting Project 2021). Although it might appear that such an approach would not be capable of significant change, numerous cities throughout the world have engaged in participatory budgeting (Ganuza and Baiocchi 2012; Sintomer et al. 2008).

This technology has several advantages. It increases transparency in local democracy (Participatory Budgeting Project 2021). It provides a way for residents to have their needs met without having to engage with bureaucracy. If they are dealing with a situation involving environmental harm, it gives locals a way to combat its ill effects (Epting 2020). In turn, if they feel anxious or afraid that a scenario is doomed without remediation, a possible avenue could exist when their elected politicians or responsible parties fail them. While this aspect might require cognitive stretching, it exhibits that possible strategies could assist in troubling matters.

Depending on the nature of participatory projects, they could qualify as modest means to combat climate change at the neighborhood level. These considerations, while not exhaustive, show that participatory budgeting could count as a saving technology. Although not all such measures will deliver positive outcomes, participatory budgeting's underlying thought pattern could lend itself to other urban planning and engineering initiatives, helping them become saving technologies (Epting 2016). This aspect should hold steady as a primary element when thinking about technologies that could help cities transition from presently wicked places to saving ones.

6.2.3   Car-Free Cities

While the novel urban technologies above could bolster saving efforts, viewing cities as technologies has the advantage of completely reconceptualizing them (or significant urban elements) could deliver outcomes that radically save the city. Car-free cities and car-free districts are examples. We must keep in mind that "car-free cities" and "car-free districts" are not merely descriptions for a place without any cars. They are thought technologies, tools to help us reconceptualize the design of urban environments. For instance, vehicle-centric transportation systems are inherently harmful in myriad ways, as examined in previous chapters, recalling their negative environmental and social impacts. It is no surprise that transportation specialists have been complaining about cars in cities for almost a century (Gordon 1931). Louis Kahn envisioned urban space without them (Richards 2016). Yet, we have reached a time wherein cities, as technologies that depend on private, fossil-fueled vehicles to connect people with places and others, kill and harm us.

This claim is not hyperbole or metaphorical. Emissions from vehicles are worsening climate change (e.g., Nam et al. 2004). Increased greenhouse gases cause respiratory-related deaths and illnesses across the globe, leading to premature deaths (Bailis et al. 2005). Children in global cities are now struggling with cognitive issues linked to traffic-related auto emissions (Cserbik et al. 2020). Tragically, a nine-year-old girl was the first person to die with auto exhaust listed as the cause of death (Laville 2020). Accepting these outcomes while doing nothing to change them meaningfully at their base entails that such situations will continue and or worsen. Aside from the apparent ethical dimensions that demand a devoted, focused study, one could make a case that people are dying so that we can continue to live for transportation (or the people profiting from it). In turn, people are standing reserve for the city's transportation system—rather than the other way.

This scenario strongly suggests that if actions do speak louder than words, leaving these conditions as they appear indicates that we have failed to develop a manner of thinking that avoids seeing the human and nonhuman worlds as nothing but standing reserve. It also suggests that if we can form a free relationship to technology and choose not to change such conditions, the horrific outcomes above are chosen. Failing to eliminate these situations—or where they stand for our urban priorities—speaks volumes. The wisdom of Dr. Martin Luther King Jr. (1963, para 11) again reminds us: "justice too long delayed is justice denied."

The benefits of such undertakings are substantial. For instance, motivated to improve air quality and provide space for pedestrians, Madrid's move to make its city center a car-free zone resulted in a 38% decrease in nitrogen oxide emissions (Medina 2019). Carbon dioxide emissions decreased by over 14% (Medina 2019). These measurable results illustrate that municipalities can take positive steps, lessening the effects of harmful wicked technologies such as too many combustion engines in a given space through applying a saving technology such as an appropriate mobility policy.

To reiterate, not every city or town would benefit from the same mitigatory measure. However, to our advantage, brave city leaders have taken bold steps toward progress, re-creating and developing the kind of actions that deliver better results, moving away from standing reserve. Dealing with these wicked technologies requires saving thinking. Following the spirit of such practices distances us from standing reserve while developing the kind of free relationships that we should want, even though this topic requires us to engage in purely normative conversations. Keep in mind that we can deliver a better mental life and sustain long-term conditions for people if we take these ideas seriously, backing them up with correlative actions.

There is a history of people championing car-free cities in the literature (McKenzie 1999; Crawford 2000; Nieuwenhuijsen & Khreis 2016). However, there are only a few real-world cases. One famous example is Venice, Italy. Built over 1500 years ago, it is car-free by default. Cars can park outside of the city space, and people can travel on narrow streets or water vehicles (Dudley 2017). City workers collect residential trash using handcarts (*ibid*). Due to the lack of combustion engines operating on the streets, the city is quiet and peaceful (*ibid*). Recalling a point from Chap. 4, this condition inherently makes a city less harmful mentally. Although this is an ancient city, it should inspire future designs that could achieve similar environmental and social outcomes.

A more recent case that *was* designed to be free of cars is Masdar City. It was envisioned as a zero-emissions city that could inform the interplay of car-free mobility and the saving city concept, but it is too early to learn that lesson. For instance, the surface areas were designed for walking, and personal rapid transit vehicles travel underground (De Graaf 2011). Yet, despite having developed a novel idea for a thoroughly planned car-free city, other elements bring significant criticism. For example, Nicolai Ouroussoff (2010, para 22–25), in his criticism of Masdar City, pinpoints why this experiment in eco-living is significantly challenged to align with a robust definition of sustainability that makes meaningful room for inclusivity:

> What Masdar really represents, in fact, is the crystallization of another global phenomenon: the growing division of the world into refined, high-end enclaves and vast formless ghettos where issues like sustainability have little immediate relevance. Ever since the notion that thoughtful planning could improve the lot of humankind died out, sometime in the 1970s, both the megarich and the educated middle classes have increasingly found solace by walling themselves off inside a variety of mini-utopias. This has involved not only the proliferation of suburban gated communities, but also the transformation of city centers in places like Paris and New York into playgrounds for tourists and the rich. Masdar is the culmination of this trend: a self-sufficient society, lifted on a pedestal and outside the reach of most of the world's citizens.

While the passage above is a bit harsh, it does highlight the reality that Masdar City's conception of "sustainability" does not meaningfully address social and economic elements that require notions of environmental and social justice that must be significantly present in sustainability's conception. Otherwise, it cannot be free and clear of over-glorified ecological performance. For instance, Julian Agyeman et al. (2002, p. 78) maintain this view: "Our interpretation of sustainability is that its focus should be to ensure a better quality of life for all and that this should be done in a just and equitable manner, whilst living within the limits of supporting ecosystems." Such views hold paramount importance. They appropriately seek to weigh approaches to sustainability so that people are not discounted or thought about as means only to secure environmentally sound ends for others—standing reserves for the "sustainable" city. We are all experiencing our anthropogenic demise of the planet together.

This point aside, the lesson here is that a saving technology must be considered related to its situatedness, meaning that one might not be optimally appreciated if its neighboring elements significantly reduce its overall effectiveness as it relates to how it can help save a city. If a city has rampant violence, pollution, and corruption, having a car-free zone or no cars does not matter much when looking at the city panoramically. Still, a takeaway should not be lost in the critique: Masdar City has developed a novel approach to navigating the cityscape. The smart move here is to study it, looking for troubled situations that might benefit from the thinking behind it, even if modified considerably.

When incorporated to examine technology, this outlook supports forming free relationships with it, necessary measures for saving cities. Considering the dire nature surrounding our global predicament, sticking with the expanded taxonomy of technology to include wicked and saving makes goal setting a manageable task. While cities are indeed complex, understanding how this sort of objective can recreate and develop them to benefit mental life and continued survival is not a sophisticated affair. To that end, focusing exclusively on understanding cities' processes fails to recall the reason for those undertakings, further revealing the danger of enframing, only seeing each other as standing reserves. If people's mental life is a peripheral consideration in the science or philosophy of the city, then we must recalibrate our methods and motivations.

6.3 Conclusion

This chapter further illustrates what saving technologies look like and how they can contribute to forming saving cities, focusing on ariel cable cars, participatory budgeting, and car-free cities/zones. While having similarities based on how they favor mental life and the long-term conditions for it, these three technologies differ in their ontological characterizations. They include physical entities and abstract devices, revealing how we can employ them to form free relationships to pursue desired goals. Showing these technologies in this manner provides a glimpse into the vast range of possibilities that remain at our disposal for pursuing measures wherein urban technologies can support humankind's collective ends that metropolitan environments can secure. While failing to act to create better mental lives for cities' residents aligns with the status quo, preserving or adding cases to such unfortunate realities.

However, the target that we must keep at the forefront of our thought is that having the ability to form a free relationship with technologies means that we can do just that. Engaging in such enterprises requires that the scope of our present inquiry must move into the normative realm. Considering the overwhelming reasons to examine transportation as identified above, such an undertaking should begin by examining the morality of urban mobility.[1] Through starting here, it reflects how we prioritize the urban problems that require immediate attention and action. While prioritizing these measures informs us about their inherently troubled states, they also tell us about the avenues that we should pursue to recreate and create them, a process that plays a role in defining us.

References

Agyeman J, Bullard R, Evans B (2002) Exploring the nexus: bringing together sustainability, environmental justice and equity. Space Polity 6(1):77–90

Bailis R, Ezzati M, Kammen D (2005) Mortality and greenhouse gas impacts of biomass and petroleum energy futures in Africa. Science 308(5718):98–103

Bullard R, Johnson G (1997) Just transportation: dismantling and class barriers to mobility. New Society Publishers, Gabriola Island

Crawford J (2000) Carfree cities. International Books, Utrecht, Netherlands

Cserbik D, Chen J, McConnell R, Berhane K, Sowell E, Schwartz J, Hackman D, Kan E, Fan C, Herting M (2020) Fine particulate matter exposure during childhood relates to hemispheric-specific differences in brain structure. Environ Int 143:105933

Dudley D (2017) The uncanny power of a city without cars. Citylab January 17. https://www.blo omberg.com/news/articles/2017-01-17/the-power-of-car-free-venice

Epting S (2016) Participatory budgeting and vertical agriculture: a thought experiment in food system reform. J Agric Environ Ethics 29(5):737–748

[1] This book is a prequel to my earlier work, *The Morality of Urban Mobility: Technology and Philosophy of the City.*

Epting S (2020) Participatory budgeting for environmental justice. Ethics Policy Environ 23(1):22–36

Epting S (2021) The morality of urban mobility: technology and philosophy of the city. Rowman Littlefield Int, London

Ganuza E, Baiocchi G (2012) The power of ambiguity: how participatory budgeting travels the globe. J Public Delib 8(2):1–12

Garsous G, Suárez-Alemán A, Serebrisky T (2019) Cable cars in urban transport: travel time savings from La Paz-El Alto (Bolivia). Transp Policy 75:171–182

Gordon C (1931) Automobiles versus railways for urban transportation. Elect Eng 50(1):4–8

De Graaf M (2011) Prt vehicle architecture and control in Masdar City. In: Automated people movers and transit systems 2011: from people movers to fully automated urban mass transit. pp 339–348

Inclusive Infrastructure (2021) Mi teleférico cable car. https://inclusiveinfra.gihub.org/case-studies/mi-teleferico-cable-car-bolivia/

King M (1963) Letter from Birmingham jail. https://www.africa.upenn.edu/Articles_Gen/Letter_Birmingham.html

Laville S (2020) Ella Kissi-Debrah: how a mother's fight for justice may help prevent other air pollution deaths. The Guardian December 16. https://www.theguardian.com/environment/2020/dec/16/ella-kissi-debrah-mother-fight-justice-air-pollution-death

Matsuyuki M, Okami S, Nakamura F, Sarmiento-Ordosgoitia I (2020) Impact of aerial cable car in low-income area in Medellín, Colombia. Trans Res Proc (48):3264-3282

McKenzie C (1999) Car-free cities-myth or possibility? Exploring the boundaries of sustainable urban transport. In: Abstracts of Volume, vol 5(1)

Medina M (2019) Emissions fall in Madrid city center thanks to new traffic restrictions. El País March 14. https://english.elpais.com/elpais/2019/03/14/inenglish/1552556189_425975.html

Menser M (2012) The participatory metropolis, or resilience requires democracy. Center Humans Nature. http://www.humansandnature.org/democracy

Moss C (2018) The most spectacular public transport system on the planet—with tickets for 33p. The Telegraph October 11. https://www.telegraph.co.uk/travel/destinations/south-america/bolivia/articles/la-paz-cable-cars/

Nam EK, Jensen TE, Wallington TJ (2004) Methane emissions from vehicles. Environ sci technol 38(7):2005–2010

Nieuwenhuijsen M, Khreis H (2016) Car free cities: pathway to healthy urban living. Environ Int 94:251–262

Ouroussoff N (2010) In Arabian desert, a sustainable city rises. The New York Times September 25. https://www.nytimes.com/2010/09/26/arts/design/26masdar.html?_r=0&pagewanted=all

Participatory Budgeting Project (2021) Participatory budgeting project. https://www.participatorybudgeting.org

Richards W (2016) Philadelphia almost: Louis Kahn's idea for a carless city. Architect Magazine. https://www.architectmagazine.com/aia-architect/aiaknowledge/philadelphia-almost_o

Sintomer Y, Herzberg C, Röcke A (2008) Participatory budgeting in Europe: potentials and challenges. Int J Urban Reg Res 32(1):164–178

Smith G (2009) Participatory budgeting: Porto Alegre. http://participedia.net/en/cases/participatory-budgeting-porto-alegre

Spickermann A, Grienitz V, Heiko A (2014) Heading towards a multimodal city of the future?: Multi-stakeholder scenarios for urban mobility. Technol Forecast Soc Chang 89:201–221

Tischler S, Mailer M (2019) Cable propelled transit systems in urban areas. Trans Res Proc 41:169–173

Wampler B (2012) Participatory Budgeting: Core principles and key impacts. J Public Delib 8(2), Article 12